A-LEVEL GEOGRAPHY TOPIC MASTER

THE WATER AND CARBON CYCLES

Series editor
Simon Oakes

Andrew Davis
Garrett Nagle

HODDER
EDUCATION
AN HACHETTE UK COMPANY

For Sue and Peter.

Acknowledgements can be found on page 228.

Photo credits

p. 6 © Garrett Nagle; **p. 11** © Garrett Nagle; **p. 21** Globaïa and Stockholm Resilience Centre; **p. 32** *t* and *b* © Garrett Nagle; **p. 47** © Garrett Nagle; **p. 59** *t* MODIS Rapid Response Team at NASA GSFC; *b* MODIS Rapid Response Team at NASA GSFC; **p. 69** © Garrett Nagle; **p. 71** © Shutterstock / Melih Cevdet Teksen; **p. 77** © Garrett Nagle; **p. 78** © Garrett Nagle; **p.81** *t* © BlackMac - stock.adobe.com; *b* © Shutterstock / Martchan; **p. 87** © Garrett Nagle; **p. 89** © Garrett Nagle; **p. 92** © sablinstanislav - stock.adobe.com; **p. 114** © allouphoto - stock.adobe.com; **p. 128** *t* © Christopher Furlong/Getty Images; *b* ©Lukasz Z - stock.adobe.com; **p. 131** *t* © Gerald Herbert/AP/REX/ Shutterstock; *b* © Milesy / Alamy Stock Photo; **p. 145** © sabangvideo - stock.adobe.com; **p. 163** *t* © Garrett Nagle; *b* © Garrett Nagle; **p. 168** © Garrett Nagle; **p. 170** © Garrett Nagle; **p. 189** *t* © stadttaube - stock.adobe.com; *b* © Newscast Online Limited / Alamy Stock Photo; **p. 190** *t* © Garrett Nagle; *b* © Garrett Nagle; **p. 206** © Garrett Nagle; **p. 208** © Vladimir Melnikov - stock.adobe.com

Orders: please contact Hachette UK Distribution, Hely Hutchinson Centre, Milton Road, Didcot, Oxfordshire, OX11 7HH. Telephone: +44 (0)1235 827827. Email education@hachette.co.uk Lines are open from 9 a.m. to 5 p.m., Monday to Friday. You can also order through our website: www.hoddereducation.co.uk

ISBN 9781510434615

© Garrett Nagle and Andrew Davis, 2018

First published in 2018 by
Hodder Education
An Hachette UK Company
Carmelite House
50 Victoria Embankment
London
EC4Y 0DZ
www.hoddereducation.co.uk

Impression number   10 9 8 7 6 5 4 3

Year   2022

Cover photo: © Helen Hotson / Alamy Stock Photo

Illustrations by Barking Dog and Aptara Inc.

Typeset in India by Aptara Inc.

Printed and bound by CPI Group (UK) Ltd, Croydon, CR0 4YY

A catalogue record for this title is available from the British Library.

MIX
Paper from
responsible sources
FSC™ C104740
FSC
www.fsc.org

Contents

Introduction

Water and carbon are essential to life on Earth and processes that link the atmosphere, land and sea. Recent reforms of the A-level curriculum reinforced the importance of physical geography by requiring study of the water and carbon cycles. This book supports and encourages learning about these biogeochemical cycles. The need to develop a thorough knowledge of the water and carbon cycles is made urgent by the challenges posed by global warming. An understanding of these cycles is central to appreciating how anthropogenic impacts are affecting the planet, both now and in the future, and in finding solutions to the problems caused by climate change.

The A-level Geography Topic Master series

The books in this series are designed to support learners who aspire to reach the highest grades. To do so requires more than learning-by-rote. Only around one-third of available marks in an A-level Geography examination are allocated to the recall of knowledge (*assessment objective 1, or AO1*). A greater proportion is reserved for higher-order cognitive tasks, including the analysis, interpretation and evaluation of geographic ideas and information (*assessment objective 2, or AO2*). Therefore, the material in this book has been purposely written and presented in ways which encourage active reading, reflection and critical thinking. The overarching aim is to help you develop the analytical and evaluative 'geo-capabilities' needed for examination success. Opportunities to practise and develop data manipulation skills are also embedded throughout the text (supporting *assessment objective 3, or AO3*).

All *Geography Topic Master* books prompt students constantly to 'think geographically'. In practice this can mean learning how to seamlessly integrate geographic concepts – including place, scale, interdependency, causality and inequality – into the way we think, argue and write. The books also take every opportunity to establish synoptic links (this means making 'bridging' connections between themes and topics). Frequent page-referencing is used to create links between different chapters and sub-topics. Additionally, numerous connections have been highlighted between *Water and carbon cycles* and other geography topics, such as *Coastal landscapes, Changing places*, or *Global systems*.

Using this book

The book may be read from cover to cover since there is a logical progression between chapters (each of which is divided into four sections). On the other hand, a chapter may be read independently whenever required as part of your school's scheme of work for this topic. A common set of features are used in each chapter:

- *Aims* establish the four main points (and sections) of each chapter
- *Key concepts* are important ideas relating either to the discipline of Geography as a whole or more specifically to the study of coastal landscapes
- *Contemporary case studies* apply geographical ideas, theories and concepts to real-world local contexts such as the response of the Amazonian rainforest to increased incidence of drought, and issues involved in managing the water and carbon-based resources at a time of climate change (e.g. the role of non-governmental and corporate organizations in tackling carbon emissions in the USA)
- *Analysis and interpretation* features help you develop the geographic skills and capabilities needed for the application of knowledge and understanding (AO2) and data manipulation (AO3)
- *Evaluating the issue* completes each chapter by discussing a key issue concerning water and carbon cycles (involving competing perspectives and views) e.g. water supply issues in the UAE and Yemen.
- Also included at the end of each chapter are the *Chapter summary, Refresher questions, Discussion activities, Fieldwork focus* (supporting the independent investigation) and selected *Further reading.*

Systems and feedback

The systems approach is a way of studying complex issues in a simplified form. It helps geographers to understand the functioning of real-world interrelationships, equilibria and feedback loops such as those associated with the water and carbon cycles. This chapter:

- explores how the systems approach is applied across a range of physical geography topics, including the water cycle, carbon cycle and studies of landscapes and rivers
- investigates the key concepts of equilibrium and feedback in a systems context
- assesses the extent to which systems change over time or always return to their equilibrium state.

KEY CONCEPTS

System An assemblage of parts and the relationships between them, which together constitute an entity or whole. The systems approach helps us visualise complex sets of interactions.

Model A simplified structuring of reality which represents supposedly significant features or relationships in a generalised form.

Feedback Ways in which changes in an environment may be accelerated or modified by the processes operating in a system.

Threshold (or tipping point) A critical level beyond which change in a system becomes potentially irreversible.

Equilibrium A state of balance between system inputs and outputs. A **steady state equilibrium** means that there is balance in the long term, although there may be short-term changes and fluctuations.

 The systems approach

▶ *What are the main characteristics of the systems approach?*

A system refers to any set of inter-related components or objects which are connected to form a working unit. A systems analysis simplifies the complexity of the natural world, and makes it more understandable. Ultimately, our entire natural environment operates as a single entity, in which each component has connections with all the other components (an idea that underpins James Lovelock's famous **Gaia hypothesis**). It is impossible to build a functioning model which takes into account all of the

 KEY TERM

Gaia hypothesis The Gaia hypothesis says that the Earth is a global control system of surface temperature, atmospheric composition and oceanic salinity. It proposes that Earth's elements (water, soil, rock, atmosphere and the living components called the biosphere) are closely integrated and form a complex interacting system that maintains the climate and biogeochemical conditions on Earth in balance (homeostasis) that best provides the conditions for life on Earth.

Subsystem A smaller part of a larger system, e.g. the basin hydrological cycle is a subsystem within the larger Earth-ocean-atmosphere system.

Drainage basin The area drained by a single river.

Earth systems shown in Table 1.1. This is why geographers identify particular environmental **subsystems** as units of study, within each of which there are plenty of clear and direct system connections that can be analysed. The drainage basin (and river landscape system within it) is one such subsystem. Chapter 2 explores how it is difficult to study the global hydrological system comprehensively, while, in contrast, studies of particular local-scale drainage basin hydrological systems can be relatively all-encompassing (with the caveat that it is of course far easier to model the behaviour of very small **drainage basins** than enormous continental basins such as the Amazon and Nile).

Systems can be represented as diagrams. In these system diagrams, stores are usually represented as rectangular boxes, and flows as arrows indicating the direction of the flow (Figure 1.1). There are several different types of system – morphological systems, cascading systems, process-response systems, open and closed systems – each of which is now explored in turn.

Morphological systems

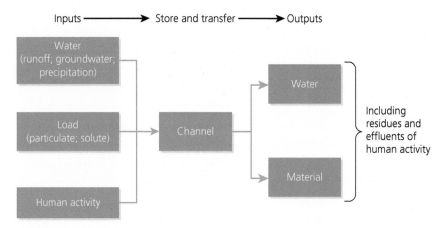

▲ **Figure 1.1** A systems diagram showing a stream channel as an open system

Morphological systems are the simplest form of system, in which some of the component parts or elements are identified, and the links between them are shown (Figure 1.1). Figure 1.1 is very much a simplification as it does not take into account either climate inputs, water inputs or human activities, for example. However, this morphological system allows geographers to investigate the functioning of certain elements within a drainage basin, and the interrelationships within it, albeit in a simplified form.

ANALYSIS AND INTERPRETATION

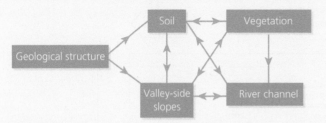

▲ **Figure 1.2** The drainage basin as a morphological system

Study Figure 1.2, which is one way of representing a drainage basin as a system.

(a) Identify the system element that has no direct influence on vegetation.

> ## GUIDANCE
>
> As you can see, geological structure is the only system element that does not *directly* link with vegetation. Instead, it is *indirectly* linked via the soil and valley-side slopes.

(b) Using Figure 1.2 and your own knowledge, suggest ways in which valley-side slopes and river channels are interrelated parts of a morphological system.

> ## GUIDANCE
>
> Interrelated things exert influence over one another:
>
> Firstly, valley slopes may have a direct and indirect influence on river channels. For example, on steep slopes, rates of overland flow may be high, and so more water and sediment will be transferred from the valley slopes to the river channel. In contrast, rates of mass movement may be much lower on low angle slopes. Valley slopes may indirectly influence river channels through their influence on soil and vegetation characteristics. Steeper slopes are more likely to have thinner soils, which in turn will influence the type of vegetation able to survive. This will influence interception, rain splash erosion, overland runoff and the volume of water and sediment reaching the river channel.
>
> Secondly, the river channel may affect valley side slopes by either steepening them through vertical or lateral erosion, or it may reduce the slope angle by depositing material at the base of the slope. In some cases, rivers may repeatedly flood onto valley slopes and begin the formation of a small floodplain, thereby burying irregularities in the valley-side slopes.

Cascading systems

A cascading system is characterised by the way *energy or matter flow through it*. The water cycle (Figure 1.3; see also page 26) and carbon cycle (see Chapter 4) – which are the main themes of this book – are both cascading systems that consist of **stores** and flows. Stores are where matter or energy are kept, and flows are the movements which provide inputs and

 KEY TERM

Stores The locations where matter or energy is held in a system.

outputs of energy and matter. The flows also involve processes that may be either transfers (a change in location) or transformations (a change in the chemical nature, a change in state or a change in energy).

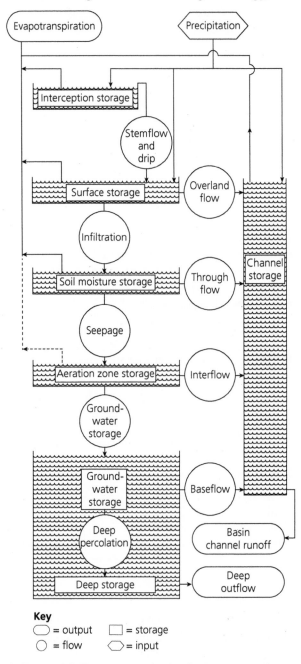

▲ **Figure 1.3** The water cycle visualised as a cascading system

Process-response systems

A process-response system integrates some of the characteristics of both morphological and cascading systems. It is used to analyse interactions between movements of energy or matter and changing environmental components. The internal dynamics of a process-response system include the operation of positive and negative feedback (see pages 12–13).

A good example of a process-response system is the drainage basin hydrological cycle, which shows inputs, stores and outputs (Figure 1.4).

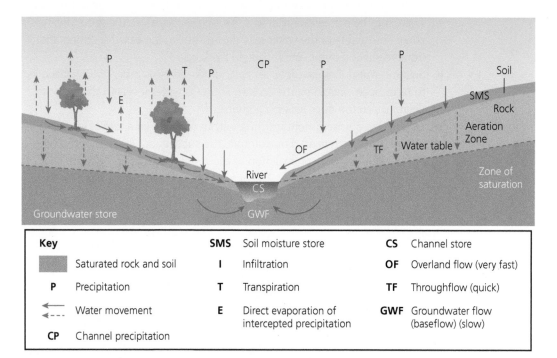

Key						
	Saturated rock and soil	**SMS**	Soil moisture store	**CS**	Channel store
P	Precipitation	**I**	Infiltration	**OF**	Overland flow (very fast)
	Water movement	**T**	Transpiration	**TF**	Throughflow (quick)
CP	Channel precipitation	**E**	Direct evaporation of intercepted precipitation	**GWF**	Groundwater flow (baseflow) (slow)

▲ **Figure 1.4** The drainage basin hydrological cycle

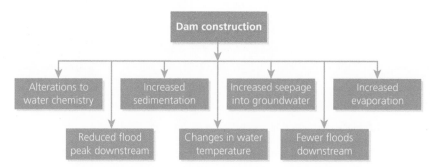

▲ **Figure 1.5** The effects of dam construction on hydrology

As Figure 1.5 shows, the hydrological and fluvio-geomorphological effects of dam construction are both intentional and unintentional. While dams may be built primarily for irrigation or flood control, there are systematic impacts resulting from the way physical processes respond to the change that has occurred, including the following:

- Increased evaporation losses due to the large reservoir contained behind the dam. For example, evaporation losses from Lake Nasser behind the Aswan Dam increased by 30 per cent following the dam's construction (it was completed in 1970).
- Thermal stratification takes place in the lake, with cold water occurring at depth and warm water at the surface. This can affect water chemistry with changes in pH levels and the amount of dissolved oxygen present.
- There may be increased water loss through seepage into the groundwater zone. The quality of the water may decline due to increased sedimentation, from vegetation dying and rotting in the lake.
- Seismic stress. For example, an earthquake in November 1981 is believed to have been caused by the Aswan Dam (as water levels in the dam decrease, so too does seismic activity).
- Deposition and sediment infilling within the lake (at a rate of about 100 million tons each year). For example, the subsequent erosion of the Nile Delta following the construction of the Aswan Dam: the Delta is shrinking at a rate of about 2.5 cm each year on account of decreased transport of new sediments.

▲ **Figure 1.6** Half-submerged trees at Batang Ai, Sarawak, Malaysia

Some of the responses listed above – notably seismic stress – involve the operation of other physical geography systems. This demonstrates the way geographers can use system theory (and process-response systems in particular) to help establish synoptic links between different interlinked geographical themes and topics. Can you think of ways in which carbon cycle flows (see pages 101–25) might also have been affected by the construction of the dam as part of a broader series of process responses?

Using the systems approach in physical geography

You will encounter many examples of system thinking while studying physical geography at A-level. Several are shown in Table 1.1.

Physical system	Possible study topics
The water cycle and hydrological systems	• Drainage basin land-use changes and their direct and indirect effects on the hydrological cycle and water flow (see page 26) • How human modification of rivers, including dam building, impacts on rivers, flooding and hydrological flows and stores (see page 30) • The impact of urbanisation on storm hydrographs (see page 49)
The carbon cycle (including soils and ecosystem studies)	• How land-use changes, including deforestation, are influencing water and carbon cycles (see page 192) • How climate change is causing carbon cycling to change (see page 122) • The impact of fire on carbon cycling in rainforests, savannas and forest systems (see page 195) • How human impact can alter succession in ecosystems
Climate systems	• Greenhouse gases and climate change (see page 20) • How El Niño, La Niña and the North Atlantic Oscillation are responsible for many short-term changes in weather (see page 16) • Changing urban microclimates (as more people live in urban areas) • Extreme weather events, including hurricane frequency and magnitude
Coastal landscape systems	• How coastal landscape systems operate • Coastal management schemes and their intentional and unintentional systematic impacts
Dryland landscape systems	• How dryland landscape systems operate • Desertification and soil degradation issues
Glacial and periglacial landscape systems	• How glacial landscape systems operate • Rising temperatures and systematic permafrost degradation and methane release (see page 206) • Rising temperatures, the melting of mountain glaciers and changing water cycle flows (page 207)
Geological and tectonic systems	• Over a long-term geological timescale, how continents move and mountains are formed or eroded • How variations in shear strength and shear resistance can trigger mass movements and landform system changes • How all drainage basins maintain a mass balance of sediments based on denudation, transport and deposition

◀ **Table 1.1** Examples of systems in physical geography

KEY TERMS

Water cycle The continuous movement of water between land, atmosphere and sea.

Deforestation The removal of trees and subsequent conversion of once forested land to non-forest use.

Carbon cycle The biogeochemical cycle by which carbon moves from one part of the global system to another. At the global scale, it is a closed system made up of linked inputs, outputs, flows and stores. At the local scale, it is an open system.

Permafrost Ground (soil or rock and included ice) that remains at or below 0°C for at least two consecutive years. The thickness of permafrost varies from less than 1 metre to more than 1.5 kilometres.

Pearson Edexcel

AQA

OCR

WJEC/Eduqas

ANALYSIS AND INTERPRETATION

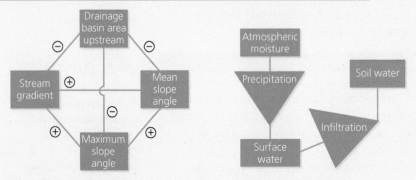

▲ **Figure 1.7a** A morphological system and **b** a cascading system

Study Figures 1.7a and 1.7b, which show two different types of system. Figure 1.7a is a morphological system (showing the relationship between slope angles and stream characteristics in a drainage basin). Figure 1.7b shows elements of the water cycle (a cascading system). The squares represent stores and the triangles transfers (flows) between stores.

(a) Analyse the relationships between the different system elements in Figure 1.7a.

GUIDANCE

When answering a question like this, remember that the question only asks for a descriptive analysis, so no explanations are required here (your analytical skills are being tested instead). Figure 1.7a shows how complicated rivers and drainage basins can be. Although it only covers four aspects of a drainage basin, there are six links between them. For example, there are positive relationships between maximum slope angle and stream gradient, mean slope angle and stream gradient and maximum slope angle and mean slope angle. In contrast, there are negative relations between drainage basin area upstream and mean slope angle, drainage basin area upstream and maximum slope angle, and between drainage basin area upstream and stream gradient.

(b) Assess the value of Figure 1.7b as a way of representing water cycle stores and flows.

GUIDANCE

Figure 1.7b shows that moisture in the atmosphere is transferred through precipitation on the Earth's surface. Some of this then enters into the soil (infiltration) to become soil moisture. The command word 'assess' requires you to look at the strengths and weaknesses of this system representation. On the positive side, it clearly distinguishes between stores and flows and shows the 'cascading' nature of the system. On the negative side, there are no relative sizes shown for the stores and flows.

(c) Compare and contrast the main features of the morphological system and the cascading system.

GUIDANCE

Both diagrams show connections between the elements of a physical system. The morphological system shows how elements interact with one another, and whether they are related in a positive or negative way. In contrast, the cascading system suggests links between some of the parts. However, it does not suggest whether the links are positive or negative. Moreover, the cascading system suggests a one-way movement of water, but not a circular movement.

Contrasting open and closed systems

Systems are normally divided into two types – open and closed. These are shown in Figures 1.8a and 1.8b (where Store A could represent moisture in the atmosphere, Store B moisture in soil/bedrock and Store C moisture in rivers/oceans).

- In a closed system, there is no transfer of energy or matter across the external boundaries of the system (Figure 1.8a). The global hydrological cycle is often considered to be a closed system, as there is a fixed supply of water always remaining within the Earth's physical system as a whole. In reality, however, the hydrological cycle is driven by solar energy, so there is a movement of energy (if not matter) across the boundary of the Earth.
- In contrast, in a truly open system, there are also flows of matter across the boundaries of the system (Figure 1.8b). In the drainage basin hydrological cycle, precipitation arrives from outside the area (for example, when a weather front arrives), and there are also movements of sediments, wind-blown material and people that all represent external inputs into the drainage basin.

▲ **Figure 1.8a** Closed and **b** open systems

ANALYSIS AND INTERPRETATION

Study Figures 1.8a and 1.8b.

(a) Using the diagrams and your own knowledge, suggest why the Earth's global water and carbon cycles are usually classed as closed systems, whereas local water and carbon cycles are open systems.

GUIDANCE

Apart from the Universe, no natural system is truly closed. Only in laboratory conditions (which are not natural) do closed systems exist on Earth. The planet is an open system with regard to energy. Energy flows from the Sun to Earth, and some is re-radiated back into space. In contrast, the Earth can be considered a closed system with regard to matter. Although in the geological past there was some input of water and materials from meteorites, the Earth generally contains all of the matter that it will ever have. Some of this, such as water, is recycled through the atmosphere and through rocks, for example. Therefore, the Earth can be considered a closed system in relation to water and carbon, for example (although both cycles are driven by solar activity, and might still be viewed as open systems with respect to incoming solar radiation).

In contrast, a drainage basin or local forest ecosystem receives energy and matter from the Sun, precipitation and higher elevations. These inputs pass through the system, performing functions such as erosion and deposition, to produce outputs of heat, water and sediment. Similarly, carbon sequestration processes taking place in a local ecosystem are balanced by carbon losses from the area (e.g. the removal of timber to other places).

Carbon sequestration
The natural capture
and storage of carbon
dioxide (CO_2) from the
atmosphere by physical or
biological processes such as
photosynthesis.

Steady state equilibrium
A long-term balance is
maintained, although there
may be short-term changes
in the system's state.

Static equilibrium A
system in which there is no
change over time.

Dynamic equilibrium
A system in which there
are short-term fluctuations
occurring over a changing
long-term mean or baseline.

Anthropogenic Relating to
human activity.

② Equilibria and feedback

▶ *Why are the concepts of equilibrium and feedback important for the study of systems?*

The key concept of equilibrium

Equilibrium refers to a state of balance between inputs and outputs. A steady state equilibrium means that there is balance in the long term, although there may be short-term changes. The precise nature of any equilibrium depends on the timescale involved.

● Over a short timescale, it may be possible to identify a static equilibrium (there is essentially no change over time) or a steady state equilibrium (there are short-term variations – such as periods of drought or flooding affecting local water cycles – over a long-term steady state).
● However, over a longer timescale the equilibrium may be dynamic (Figure 1.8), meaning that short-term fluctuations *are additionally occurring around a changing long-term mean* (linked, for example, with progressive natural or anthropogenic climate change).

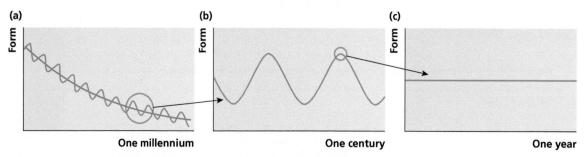

(a) Form — One millennium
(b) Form — One century
(c) Form — One year

▲ **Figure 1.9** Timescales and equilibria: **a** dynamic equilibrium **b** steady state equilibrium and **c** static equilibrium

Over a short-term period, for example weeks or months, no change may be visible, and equilibrium could be said to be static. A steady state equilibrium is said to exist when there are minor changes to a system, but it always returns to its original form (possible because negative feedback occurs, as page 12 explains). Over a longer term, change may be visible, and dynamic equilibrium is said to occur; i.e. the whole system is very gradually changing due to changes in the wider environment, such as long-term tectonic plate movement affecting a location's altitude and/or climatic characteristics. A good example of this is the continued gradual isostatic uplift of Scottish beaches and the gradual sinking of the south-east of England following the last major glacial advance over 110,000 to 11,700 years ago. Because of these processes, UK coastal landscape systems are in a state of dynamic equilibrium when viewed over millennia.

The different kinds of equilibria shown in Figure 1.9 can also be illustrated using changes in stream discharge.

- Mean river flow generally stays the same for a period of days (Figure 1.9c). However, following a storm, stream flow (discharge) may increase over the short term. After a few days or longer – depending on the extent of the storm and size of river basin – stream flow returns to 'normal'. There may also be seasonal variations in discharge in some climatic zones: the year-on-year pattern generally remains the same: i.e. there is still a steady state equilibrium.
- Longer-term cyclical variations lasting for years and decades may occur; these might be associated with El Niño events (see page 16) or longer-lasting climatic oscillations. But even these changes will usually be reversed, allowing conditions to return to 'normal' after some years (Figure 1.9b).
- However, over a long-term timescale of millennia, climates can permanently alter, along with stream flows (Figure 1.9a). For example, many streams in southern England had much higher flows during the most recent periglacial climatic phase around 9,000 years ago – with discharges up to 50 times greater than today during periods of spring snowmelt from retreating ice sheets (Figure 1.10). Subsequent changes leading to the climate we experience in the UK today are part of this long-term dynamic trend.

Short-term misconceptions about long-term changes

Humans have only been measuring natural phenomena for a relatively short period of the Earth's history and this can make it difficult to identify long-term trends and the equilibria states (or otherwise) of different systems. Reliable long-term data are required to identify long-term trends and to differentiate them from short-term fluctuations. However, long-term data may not always be available and so trends or patterns may not be correctly identified i.e. different trends occur over different time scales/sampling frequencies.

- For example, Figure 1.11a suggests that a researcher measuring an environmental change (such as atmospheric carbon) from period t1 to t2 would most likely conclude that the level was falling; whereas there is, in fact, a steady state when the system is viewed over a longer time period.
- In Figure 1.11b, a researcher studying the period lasting from t1 to t2 might conclude the system was in a steady state; but if he or she had access to data over a longer time period it would become apparent that the system is actually in a state of dynamic equilibrium.

As many processes that operate in physical geography do so over a very long period of time, it is important, but sometimes not possible, to have access to quantitative data over long periods of time. Unfortunately, some processes, such as glacial erosion, may remove the evidence from earlier periods, making it difficult to build up a clear picture of landscape development.

▲ **Figure 1.10** At the end of the last ice age, glaciers in the UK melted gradually over many centuries: this introduced an additional element of long-term change, or dynamic equilibrium, to the seasonal flow variations experienced by rivers

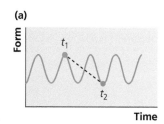

(a)

Form

t_1

t_2

Time

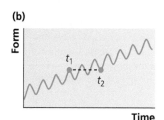

(b)

Form

t_1

t_2

Time

▲ **Figure 1.11** Short-term misconceptions (in relation to long-term trends)

Dynamic metastable equilibrium

Finally, some geographers have noted that there may be episodes of landscape stability followed by periods of landscape change – this is known as 'episodic erosion'. Stanley Schumm (1976) showed that some valley floors are lowered episodically rather than continuously. This could occur due to the accumulation of sediments from the upper part of a basin covering, and protecting, the bedrock on the valley floor. Similarly, the build-up of carbon (locked in frozen soil) in periglacial areas over centuries could be removed following warming conditions and the occurrence of fires, thereby releasing vast amounts of carbon and methane into the atmosphere.

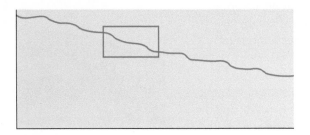

This situation has been described as **dynamic metastable equilibrium** (Figure 1.12), which differs from dynamic equilibrium in that the landscape is lowered discontinuously rather than progressively. Under dynamic metastable equilibrium, phases of steady state equilibrium are altered when an environmental threshold has been exceeded.

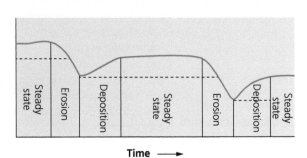

Time ⟶

▲ **Figure 1.12** Episodic erosion and dynamic metastable equilibrium

Feedback in systems

One important characteristic of systems is the way they sometimes self-adjust in response to throughputs of energy and matter. This change is brought about by a self-regulation mechanism called feedback. As one of the components in a system changes, an imbalance may result. As a result, the whole system begins to change in ways which accommodate the changed component. The time-lag between external change and internal adjustment is known as the relaxation time of the system.

Negative feedback

The most common type of feedback relationship is called **negative feedback**. This occurs when a change factor leads to a number of other linked changes, but eventually the whole system stabilises again, and a steady state equilibrium is restored.

For example, an increase in rainfall may lead to an increase in stream velocity, which causes greater erosion and an increase in stream width. However, this now results in an increased wetted perimeter and proportionately greater friction between water and the river bed. Stream velocity therefore falls back to its original value. This kind of negative feedback is very common in physical geography. Note, however, that multiple events and responses – occurring over a long geological time period – may also result in the dynamic metastable equilibrium sequence shown in Figure 1.12.

Figure 1.13 shows an example of negative feedback or self-regulation in a landscape system. The initial trigger for change might be tectonic movement, causing uplift of rocks, or faulting, which results in some rocks being raised relative to others. If uplift across a river bed should cause an increase in channel gradient, the resulting increase in stream velocity will most likely cause an increase in erosion and vertical down-cutting into the bedrock. In time, however, this could lead to a reduction in the stream's gradient again: equilibrium has been restored.

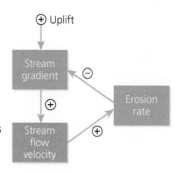

▲ **Figure 1.13** Negative feedback can help regulate water cycle flows and processes in a river channel

Positive feedback

In contrast, positive feedback, which leads to a greater, permanent deviation from the original condition, is rarer. Initial changes are amplified (rather than being cancelled out) by internal system processes.

- Positive feedback involves the crossing of a threshold that can lead to a system abandoning its old steady state equilibrium altogether. For example, increased temperature through global warming melts more of the ice in polar ice caps, glaciers and sea ice, leading to a decrease in the Earth's reflectivity (albedo); thus the Earth absorbs more of the Sun's energy, which makes the temperature increase even more, melting more ice.
- How long it takes for a new (but different) steady state equilibrium to be achieved (if ever) depends very much on the context involved. IPCC scientists are alarmed by some climate change scenarios precisely because they appear to suggest ever-accelerating change.

(Positive feedback and climate change are explored in greater depth on page 146.)

Figure 1.14 shows an example of positive feedback occurring in a local water cycle context. In this case, rainfall and runoff from a severe storm has eroded the uppermost layer of a drainage basin's soil. This layer is generally more permeable than the lower layers of the soil. As the exposed lower layers cannot soak up the rain collecting on its surface, the amount of water flowing over the land increases, and so too does soil erosion. This may continue until all of the erodible sub-soil has been removed and the underlying bedrock has been exposed. The drainage basin's soil – an important water and carbon store (see pages 40–1 and 115–16) – has been permanently destroyed, with lasting implications for local water and carbon cycling.

Positive feedback in the natural environment usually operates in short episodes of destructive activity but, in the longer term, negative feedback and self-regulation tend to prevail.

🔑 **KEY TERMS**

Positive feedback A change in a system which leads to an increasing deviation from the original (also known as vicious circle, snowball effect).

Threshold (or tipping point) A stage in a system's development when it irreversibly departs from a previously established steady state equilibrium.

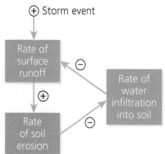

▲ **Figure 1.14** Positive feedback in the water cycle following a storm event

KEY TERMS

Climate change A change in global or regional climate patterns, caused largely by increased levels of atmospheric carbon dioxide produced by the use of fossil fuels.

Flow A movement (or transfer or flux) between stores in a system.

Input A flow entering a store.

Output A flow leaving a store.

Using models in physical geography

System definition is a key part of model design, and helps us boil systems down to their component parts (simplifying the complexity), and is a core part of the scientific method. A model is a simplified version of reality which is used to improve our understanding of how systems work and can help predict how they will respond to change. Computer models use current and past data to generate future predictions. All models have strengths and limitations and inevitably involve some simplification and loss of accuracy. Some models are complex, such as the computer models that predict the effect of climate change (see page 15). Diagrammatic models (such as the water cycle on page 4 and the carbon cycle on page 101) can help an audience visualise the flows, stores and linkages that make up complex systems.

There are many advantages of using models. They help scientists make predictions of what will happen if there are changes to system inputs, outputs or storages. Moreover, models allow inputs to be changed and outcomes examined without having to wait a long time (as we would have to if studying real events). In addition, they allow results to be shown to other scientists and to the public, and can be easier to understand than detailed information about the whole system. Building physically based computer models is the established scientific method for studying global circulation and underpins most change science.

However, models have limitations too. For example, environmental factors can be very complex with many interrelated components; it may be therefore impossible to take all variables fully into account. Unfortunately, if models are used to provide an overly simplified representation of reality then they may become less accurate as predictive tools. For example, there are a great many complex factors involved in the operation of atmospheric systems and carbon cycle fluxes (see page 102). As a result, some climate change sceptics have criticised the models used by the Intergovernmental Panel on Climate Change (IPCC) on the grounds of validity (their argument being that the processes and issues are so complex that any attempt at modelling is likely to result in potentially oversimplified and therefore flawed findings).

Different models may show varying effects and outcomes despite using the same input data. For example, competing models used to predict the effect of climate change can provide contrasting results, and levels of uncertainty tend to increase the further into the future we look. Moreover, any model is only as good as the data inputs used, and some information may not be reliable in the first place. Models also rely on the expertise and impartiality of the people making them. Different people may interpret models in varying ways, however, and so come to different conclusions. People who would gain from the results of particular models or predictions may exhibit bias by using them to their own advantage. For this reason it is very important that the workings of a model are understood and assessed. Scientific consensus is developed through the process of peer review whereby independent scientists assess the validity of other groups' work.

Figure 1.1 shows long-term climatic scenarios described by the IPCC. The two climate change predictions portrayed here are worst-case RCP8.5 and

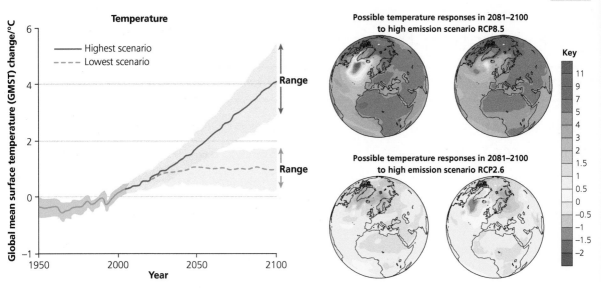

▲ **Figure 1.15** Long-term climatic uncertainties

best-case RCP2.6. The globes in Figure 1.1 show the likely ranges for regional warming by 2100. The models agree on large-scale warming at the surface, and that the land is warming faster than the ocean and that the Arctic will warm faster than the Tropics. Overall, however, they differ greatly in what they show, and the further into the future they look, the greater they diverge.

Moreover, there are two levels of uncertainty shown here: in addition to the two competing high-emissions and low-emissions scenarios produced by climate change modellers, there is further uncertainty surrounding *exactly* how much warming will occur for both scenarios. This is because the climate system and global economic systems responsible for carbon emissions are both so complex. For example, climate change will depend, in part, on the emissions of greenhouse gases from emerging economies and developed countries. Even if the exact volume of future greenhouse gases could be calculated reliably by climate change modellers, there may still be unknown feedback mechanisms and tipping points (see page 12) that will affect outcomes in unanticipated ways.

Trying to predict human behaviour accurately in advance over the next 80 years or so is in any case clearly impossible. Future climate change will be affected by economic development processes, population growth trends, technological advances (such as the 'green' technologies analysed on pages 156–86), political decisions (to develop either renewable or non-renewable energy sources) and the varying ability of some countries to pay for whatever changes are required. With all these factors in mind (and there are many more), the IPCC develops different predictions to take into account changes that may influence climate change. They produce a set of alternative models to take account of the range of possibilities.

Finally, it is worth noting that Figure 1.15 shows larger-scale average changes, such as mean global temperature changes and broad regional patterns of warming. Far more detailed models are needed to show, for example, local precipitation changes for particular towns or cities.

🔑 **KEY TERM**

Greenhouse gas
Atmospheric gases, such as carbon dioxide, allow short-wave radiation from the Sun to pass through them, but they trap outgoing long-wave radiation, thereby raising the temperature of the lower atmosphere.

▶ *To what extent do different physical systems change over time in permanent ways?*

Possible contexts for systems change

The Earth-atmosphere-ocean system has many subsystems operating at a variety of scales. For example, some lithospheric and tectonic systems operate very slowly and may appear unchanged for millennia. The current distribution of continents is a good example (albeit with very small movements measured in just centimetres per year). Similarly, ice stored on Antarctica – which is several kilometres thick in places – is thought to have been present for at least 2 million years. Some scientists believe it could be up to 40 million years old.

In contrast, the state of the atmosphere is constantly changing. It varies day by day, season by season, and over a period of decades, centuries and millennia. These different timescales for change are all operating simultaneously. For example:

- in some mountainous areas, there may be frequent diurnal changes in surface water, soil water and some small streams from ice to water; as a result there are daily cycles of change in local hydrological and landscape systems that can be studied
- some ecosystems show clear seasonal changes in vegetation cover and productivity – temperate deciduous woodland and savannas are good examples, whereas tropical rainforests and coral reefs show little seasonal variation
- even where there is marked seasonal ecosystem (and thus carbon cycling) change, however, a steady state equilibrium is most likely maintained for the entire year as a whole. Any debate over whether change is occurring or not therefore becomes a matter of temporal perspective!

Evaluating the view that some systems *do not* change over time in permanent ways

Certain regions of the Earth and its surface features have remained essentially the same for very long periods of time. Antarctica and the tropical rainforest have already been mentioned, but some of the world's great mountain landform systems, such as the Himalayas, have also existed in broadly their current state for many millennia. Very dry conditions persist throughout the world's deserts, and despite occasional rains accompanied by brief flurries of vegetative growth, they return shortly afterwards to being dry, bare surfaces. This can be viewed as another example of very long-term steady state equilibrium. Other processes may be more frequent, and longer lasting, but return, nevertheless, to a steady state equilibrium (for example, ecosystem regeneration following the impact of a hurricane or a volcanic eruption).

The El Niño Southern Oscillation (ENSO) provides one further example of equilibrium being broadly maintained over time – despite the regular occurrence of shorter-term climatic fluctuations associated with this atmospheric phenomenon. ENSO is a reversal of the normal air and ocean circulation pattern for low latitudes. During El Niño years, a periodic warming of the eastern Pacific occurs, lasting for two years and occurring at intervals between two and ten years.

The normal east–west circulation that El Niño interrupts is called the Walker loop or Walker circulation (Figure 1.16). Near South America, under normal conditions winds blow offshore,

A normal year

Warm, moist air rises, cools and condenses, forming rain clouds

Walker loop – normal air circulation

The air pushes the warm water westwards

In a non-El Niño year, the trade winds blow from east to west along the Equator

(a)

Rainforests

Australia

Warm water

Warm water

Cold water

Trade winds

Thermocline

Upwelling

Deserts

(1) The trade winds blow Equator-wise and westwards across the tropical Pacific
(2) The winds blow towards the warm water of the western Pacific
(3) Convectional uplift occurs as the water heats the atmosphere
(4) The trade winds push the warm air westwards. Along the east coast of Peru, the shallow position of the thermocline allows winds to pull up water from below
(5) This causes upwelling of nutrient-rich cold water, leading to optimum fishing conditions
(6) The pressure of the trade winds results in sea levels in Australasia being 50 cm higher than Peru and sea temperatures being 8°C higher
(7) The Walker loop returns air to the eastern tropical Pacific

An El Niño year

Air circulation loop reversed

Disrupted trade winds

(b)

Warm water reversal

Warm water

Cold water

Disrupted trade winds

The trade winds pattern is disrupted – it may slacken or even reverse and this has a knock-on effect on the ocean currents

(1) The trade winds in the western Pacific weaken and die
(2) There may even be a reverse direction of flow
(3) The piled-up water in the west moves back east, leading to a 30 cm rise in sea level in Peru
(4) The region of rising air moves east with the associated convectional uplift. Upper air disturbances distort the path of jet steams, which can lead to teleconnections all around the world
(5) The eastern Pacific Ocean becomes 6–8°C warmer. The El Niño effect overrides the cold northbound Humboldt Current, thus breaking the food chain. Lack of phytoplankton results in a reduction in fish numbers, which in turn affects fish-eating birds on the Galapagos Islands
(6) Conditions are calmer across the whole Pacific

A La Niña year

Very strong Walker loop

(c)

Very warm water

Warm water

Cold water

This an exaggerated version of a normal year, with a strong Walker loop.
(1) Extremely strong trade winds
(2) The trade winds push warm water westwards, giving a sea level up to 1 m higher in Indonesia and the Philippines
(3) Low pressure develops with very strong convectional uplift as very warm water heats the atmosphere. This leads to heavy rain in south East Asia
(4) Increase in the equatorial undercurrent and very strong upwelling of cold water off Peru results in strong high pressure and extreme drought. This can be a major problem in the already semi-arid areas of northern Chile and Peru

▲ **Figure 1.16** Normal conditions in the South Pacific Ocean are shown in **(a)**, El Niño conditions are shown in **(b)** and La Niña conditions are shown in **(c)**

causing upwelling of the cold, rich waters. By contrast, warm surface water is pushed into the western Pacific. Normally sea surface temperatures (SSTs) in the western Pacific are over 28°C, creating an area of low pressure and producing high rainfall. By contrast, over coastal South America, SSTs are lower, high pressure exists and conditions are dry.

During El Niño episodes, this pattern is reversed. Water temperatures in the eastern Pacific rise as warm water from the western Pacific flows eastwards. Low pressure develops over the eastern Pacific as water temperature rises, while high pressure takes hold over the western Pacific. Consequently, heavy rainfall occurs over coastal South America; in contrast,

Indonesia and the western Pacific are now warm and dry. Not all El Niño events are the same. There are stronger and weaker events, which may be linked to climate change.

In some other years, a phenomenon called La Niña occurs instead. This is an intensification of the normal Walker circulation whereby strong easterly winds push cold upwelling water off the coast of South America into the western Pacific. Its impact extends beyond the Pacific and has been linked with unusual rainfall patterns in the Sahel (just to the south of the Sahara desert) and in India, and with colder and wetter condition in western Canada.

A view can therefore be formed that the climate system in the southern Pacific region is essentially stable and unchanging *when a longer-term analysis of air and ocean circulation is made.* Periodic El Niño and La Niña reversals in circulation are part of the system's normal steady state. Moreover, local water cycles in those parts of South America most affected by ENSO cycles enjoy a steady state in the longer term, despite alternating years or longer periods of very low and very high rainfall: water stores that are exhausted during La Niña years are replenished during El Niño years (see also page 122).

Evaluating the view that some systems *do* change in permanent ways over time

When we study physical systems over a longer timescale of millennia, or even millions of years, there is much greater evidence for significant and permanent changes taking place. There have been many ice ages during the Earth's geological history:

- The earliest ice age was some 2.3 billion years ago and the current ice age is in a warm interglacial phase with ice largely occurring in high latitudes and altitudes. However, much of this ice currently shows signs of receding. Currently, about 10 per cent of the Earth's surface is covered by ice, but as recently as 18,000 years ago, the figure was 30 per cent.
- Geologists also believe that the Earth may have become a giant 'snowball' during two distinct Cryogenian ice ages which occurred around 650 to 750 million years ago.

There are a number of interrelated factors that may cause ice ages and glacial phases. These include, firstly, the 'stretch' in the Earth's orbit around the Sun, the 'tilt' of the Earth and the 'wobble' of the Earth's axis (Figure 1.17). Secondly, the distribution of land and sea has changed radically over time in response to tectonic activity. Antarctica's arrival at the South Pole led to the formation of the Antarctic ice sheet; this brought about permanent and fundamental changes in the Earth's climate system and global water cycle.

Also, the collision of tectonic plates has resulted in mountain-building processes. The Himalayas are still rising today, part of an uplift initiated over 30 million years ago. The formation of the Himalayas was important for the development of the monsoon system; the new mountains' great mass forced air to rise, condense and produce rainfall (in contrast, where there is a lack of high ground, such as in western Pakistan and the Thar Desert, the monsoon has little effect). Such a major change in South Asia's relief had major influences on water cycle system operations. Without the Himalayas, northern India would most likely be a cold desert today; but because of the Himalayas, rainfall is increased and cold winds from Siberia are checked. This change also triggered a cascade of changes for the carbon cycle: a great thickness of uplifted sedimentary rocks, such as limestone, has been eroded over time by monsoon rains, leading to an increase in the volume of carbon transferred in solution into the sedimentary reservoir (ocean sediments).

The 100,000-year stretch
The Earth's orbit changes from being relatively circular to having an elliptical shape and back again in a cycle that takes 100,000 years. Glacials occur when the orbit is relatively circular, and interglacials when it has a more elliptical shape.

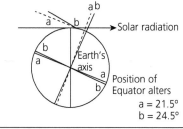

Earth – elliptical orbit
Sun
Earth – more circular orbit

The 42,000-year tilt
Although the Tropics of Cancer and Capricorn are currently located at 23.5°N and S respectively, the earth's axis varies between 21.5° and 24.5°. When the tilt increases, summers become hotter and winters colder, leading to conditions that favour interglacials.

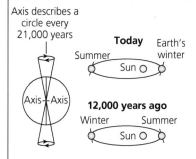

Solar radiation
Earth's axis
Position of Equator alters
a = 21.5°
b = 24.5°

The 21,000-year wobble
The Earth slowly wobbles in space over a 21,000-year period. At present, the wobble places the Earth closest to the Sun in the northern hemisphere's winter and furthest away during the northern hemisphere's summer. This tends to make winters mild and summers cool. These are ideal conditions for glacials to develop. The position was in reverse 12,000 years ago, and this contributed to the warm interglacial.

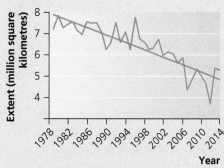

Axis describes a circle every 21,000 years

Today
Summer
Earth's winter
Sun

Axis — Axis

12,000 years ago
Winter
Summer
Sun

▲ **Figure 1.17** The Earth's stretch, tilt and wobble

Projected permanent changes in Arctic sea ice coverage

Sea ice is part of the hydrological cycle in the Arctic Ocean. However, in recent decades the extent and thickness of the ice has been reduced, and this has been linked with changes in the system caused by human activities, notably the increase in greenhouse gases, and their impact on the cryosphere. Sea ice is generally thinner than terrestrial ice, and so is more vulnerable to melting.

Arctic sea ice has declined dramatically since about 1975 (Figure 1.18). The main reason is believed to be global warming, although this is not the only reason. The sea ice recorded on 3 February 2016 was the lowest volume on record. The Arctic is believed to now be at its warmest for 40,000 years, and the length of the melting season has increased by nearly three weeks since 1979.

▲ **Figure 1.18** The extent of Arctic sea ice, 1979–2014

The Arctic sea ice minimum is generally reached during September and the maximum during March. However, the overall volume, thickness and extent have been declining for decades. In addition, the time the ice remains is changing. For example, in 1988, ice that was more than four years old accounted for over one-quarter of the Arctic sea ice, but by 2013 it was less than one-twelfth.

As the ice has receded, the potential for wave formation has increased. In 2012, five-metre waves were recorded in the Beaufort Sea. These waves helped break up the sea ice, thus establishing a positive feedback loop of disappearing sea ice and wave formation. Under 'normal' conditions, sea ice prevents the formation of waves.

Scientists have debated when (not if) the Arctic will become ice-free during summer. Predictions range from 2016 to 2040. This is viewed as a very long-term change that will soon occur in the absence of substantial climate change mitigation measures (see also pages 156–86).

The Anthropocene: a new epoch of changing systems?

The Quaternary, which existed for about 2 million years, was distinguished by regular shifts into and out of glacial and interglacial phases. The most recent geological epoch – the Holocene – has been a relatively stable part of the Quaternary period. However, scientists believe that we have entered a new epoch, and have called it the Anthropocene. They believe that this is a permanent change from the Holocene system, and that this new system is dominated by human activities.

- One of the main ways in which humans have affected the planet is through the burning of fossil fuels. In the last 200 years, people have released quantities of fossil carbon that the planet took hundreds of millions of years to store away. This has led to major changes in the Earth's carbon cycle. Although the natural fluxes of carbon dioxide into and out of the atmosphere are still more than ten times larger than the amount that humans put in every year, the human addition matters disproportionately because it unbalances those natural flows.
- The result of putting more carbon into the atmosphere is a warmer climate, a melting Arctic, higher sea levels, an increase of evaporation and precipitation and new ocean chemistry (see page 108). Some climate scientists argue that the goal of the twenty-first century should be not just to stop the amount of carbon in the atmosphere increasing, but to start actively decreasing it. This might be done in part by growing forests, but it might also need more high-tech geoengineering, such as burning newly grown plant matter in power stations and pumping the resulting carbon dioxide into aquifers.

Another characteristic of the Anthropocene is changes in most ecosystems on the planet that reflect the presence of people. Farmland covers about 50 per cent of the world's surface, up from about 5 per cent in 1750. Almost 90 per cent of the world's plant activity is believed to be found in ecosystems in which humans play a significant role. Humans have not just spread over the planet, they have changed the way it works, including the functioning of the climate system, global and local water and carbon cycles and some landform systems, including managed coastlines, retreating glaciers (on account of a warming atmosphere) and river sediment movements.

The 'planetary boundaries' view of system change

Planetary boundaries are the limits for the processes that determine the Earth's capacity for self-regulation. Recent evidence suggests the planetary system as a whole is approaching its boundary limit and that harmful changes now occurring may be permanent. Atmospheric concentrations of carbon dioxide are now in excess of 400 ppmv (parts per million by volume). The loss of summer polar sea ice is almost irreversible and will have a number of feedback mechanisms that will make the Earth even warmer and the sea level higher than at present. The consequences of human modification of water bodies include global-scale river-flow changes and land-use changes, causing changes in evaporation and runoff. This is likely to lead to permanent changes in the water cycle, and the carbon cycle, with unknown consequences.

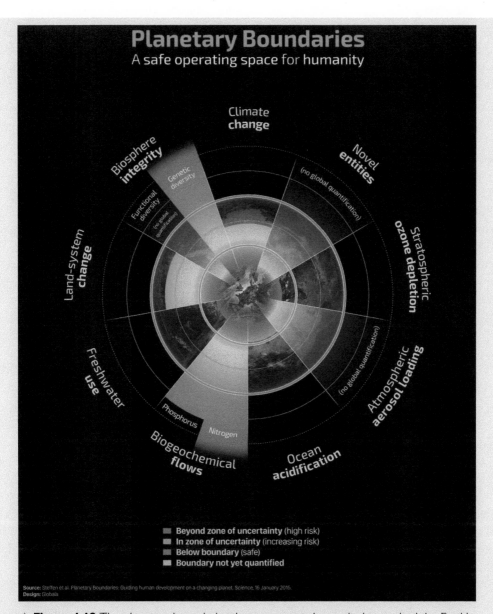

▲ **Figure 1.19** The planetary boundaries theory suggests humanity has pushed the Earth's physical systems beyond a threshold of permanent and harmful changes (Steffen et al; 2015)

Climate change and biosphere integrity are called 'core boundaries'. Altering these – as we are doing – is driving the Earth system into a new, less hospitable system state.

Alternative viewpoints on the permanency of physical systems changes

There are, however, some scientists who reject the view that the world has irreversibly entered a new negative era dominated by human activity. They say that solar energy remains the driving force for many of the Earth's systems, such as climate and weather. They point out that human activity has had very little, if any, effect on tectonic systems.

An extreme version of this strand of thought is represented by climate change sceptics, who continue to deny that changes in the climate are anything other than part of the natural cycle, meaning that temperatures could fall again in time (especially, it is argued, if a new glacial period begins).

- *The Sceptical Environmentalist* was written by the Danish environmentalist Bjørn Lomborg. In this book, he argues that many system changes such as global warming, biodiversity loss and water shortages are unsupported by statistical analysis.
- *The Great Global Warming Swindle* by Martin Durkin is a documentary that argues against the consensus scientific view that changes in atmospheric systems are likely to be due to an increase in anthropogenic emissions of greenhouse gases. Durkin thinks that increased sunspot activity could be responsible for increases in temperature.

These views are rejected by the overwhelming majority of the scientific community.

Arriving at an evidenced conclusion

There is no doubting that the Earth's many physical systems, including the global water and carbon cycles, are constantly changing. Some changes occur daily, others are seasonal or decadal; others may be operating over hundreds of millions of years. To what extent the changes we analyse are actually fluctuations occurring naturally within a long-term steady state equilibrium – or point instead to a longer-term dynamic (changing) equilibrium, or dynamic metastable equilibrium – can often become a moot point on account of data limitations or competing interpretations of the same data.

There are certain reasons to suggest that, even if the Earth's physical systems have been stable for the last few hundreds of thousands of years, the planet is now experiencing accelerating and possibly irreversible change. A new geological epoch, the Anthropocene, has been proposed (in which, increasingly, the footprint of human activity appears in all systems). The world's population is larger than it has ever been, and the demands we make on the environment are increasing rapidly. There are few parts of the Earth's surface that have not been affected by human impact, either directly or indirectly. And, as more people reach a higher standard of living, there could be changes to lifestyles that have an even greater impact on physical systems, especially the water and carbon cycles. Some scientists have even gone so far as to establish planetary boundaries, beyond which they say there may be irreversible changes occurring in multiple systems.

Nevertheless, it may be possible that some time in the future, humanity develops 'technological fixes', such as a commercially affordable alternative to fossil fuels, *in vitro* farming (using stem cells to produce meat) or widespread desalination of water. These and other developments could slow down the human impact on the planet's physical systems. But they have not become widespread yet.

🔑 KEY TERMS

Lithospheric system The inputs of energy and matter into the lithosphere (crust and uppermost part of the mantle), the processes operating there, including plate tectonics, denudation, transport and the outputs of energy and matter.

Ecosystem A community and the physical environment with which it interacts.

Glacial phase Cold phases during an ice age, when ice masses increase in volume.

Interglacial phase Warm phases during an ice age when ice masses decrease in volume/size.

Cryosphere Those areas of the Earth where water is frozen into snow or ice, including ice sheets, ice caps, alpine glaciers, sea ice and permafrost.

Fossil fuel Non-renewable carbon resources, including oil, coal, natural gas and shale gas.

Biosphere The areas of the Earth that contain life.

Chapter summary

✔ The systems analysis approach is a commonly used way of studying complex phenomena.

✔ Geographers study systems ranging in size from the Earth as a whole to local drainage basins and even smaller physical features.

✔ The water cycle and carbon cycle are two systems universally studied in A-level Geography at both global and local scales.

✔ There are different types of systems, including: morphological systems; cascading systems; process-response systems; open and closed systems.

✔ Equilibrium refers to the state of balance between inputs and outputs; a steady state equilibrium means that there is balance in the long term, although there may be short-term changes.

✔ Systems may change over time – some changes are more continuous (dynamic equilibrium) whereas others are more episodic (dynamic metastable equilibrium).

✔ System changes can occur over any timescale, from very short-term (such as diurnal variations in weather), to seasonal, annual, decadal and even geological time (millennia).

✔ The threshold (or tipping point) concept deals with a critical level of system change beyond which change is irreversible on account of often unpredictable positive feedback mechanisms.

✔ Some scientists have suggested that the Earth is entering a new geological era, the Anthropocene, due to the impact of human activity now exceeding the planetary boundaries (thresholds) of different physical systems.

Refresher questions

1 Outline the advantages and disadvantages of relying on models in geography.

2 Using examples, distinguish between morphological systems and cascading systems.

3 What is meant by the following terms: open system; closed system; threshold?

4 Using examples, distinguish between steady state equilibrium, dynamic equilibrium and dynamic metastable equilibrium. Why might a researcher struggle to identify which state a particular system is in?

5 Briefly explain why the global hydrological cycle is generally treated as a closed system whereas the basin hydrological cycle is viewed as an open system.

6 Using examples, explain how negative feedback and positive feedback processes can affect the long-term operation of different systems.

7 Outline the evidence that suggests that we are now living in the Anthropocene, a new geological epoch.

8 Using evidence, suggest why some geographers reject the Anthropocene concept.

Discussion activities

1   Discuss, design and justify a systems diagram that shows your school or college in terms of inputs, processes and outputs. This can be done as a whole class activity or in small groups.

2   In pairs, discuss the challenges researchers face when investigating whether changes affecting a system are permanent or part of a longer-term steady state equilibrium (maintained by negative feedback). What information would help a researcher understand more about the situation?

3   Study Figure 1.12, which shows dynamic metastable equilibrium for a river valley. In a small group, design an annotated diagram(s) to show processes operating on a coastal cliff or glacial valley wall that has experienced a major landslide. Explain how the area may experience dynamic metastable equilibrium over a long period of time.

4   Take a close look at your local school or college environment. Identify ways in which the environment has been modified by human activity at different geographic scales (for example, by local people, national government or global influences). Discuss the extent to which these activities support the idea of the Anthropocene era.

5   In pairs, discuss the value of the planetary boundaries concept. What practical uses could it serve?

Further reading

Barry, R. G., and Chorley, R. (2009) *Atmosphere, Weather and Climate*, Routledge

Goudie, A. S. (1977) *Environmental Change*, Oxford

Holden, J. (ed.) (2012) *An Introduction to Physical Geography and the Environment*, Pearson

Peterson, J., *et al.* (2014) *Fundamentals of Physical Geography*, Cengage Learning

Steffen, W., *et al.* (2015) 'Planetary boundaries: guiding human development on a changing planet', *Science* 347, 1259855

Summerfield, M. (1991) *Global Geomorphology*, Longman

Vince, G. (2014) *Adventures in the Anthropocene*, Vintage

CHAPTER 2

Water cycle dynamics

The water cycle is a dynamic system operating at multiple spatial and temporal scales. Increasingly, the cycle is influenced by human activities, both directly and indirectly. This chapter investigates how natural processes and stores operate, and how human activities affect them. This chapter:

- examines the global hydrological cycle and human influences
- investigates drainage basin processes and the effect of human impacts
- explores seasonal river regimes and short-term flood hydrographs
- assesses the variability of system flows over time.

KEY CONCEPTS

Water balance The water balance (or water budget) of an area over a period of time refers to the way in which precipitation is partitioned between the processes of runoff, evapotranspiration and various forms of storage.

Temporal variability Most rivers exhibit variability over a number of timescales. On a short timescale (i.e. less than a week), a storm or flood hydrograph shows how a river responds to a particular storm. Over the course of a year, a river regime shows how a river responds to seasonal changes in rainfall, temperature and evapotranspiration.

 The global water cycle and its stores

▶ *What are the main processes and stores in the global hydrological cycle?*

The global hydrological cycle involves the perpetual movement of water between atmosphere, lithosphere and biosphere (Figure 2.1). Although the global hydrological system could be viewed as an open system due to the input of solar energy which drives its flows and fluxes, it is generally thought of as a closed system because there are no losses or gains of water over time (see page 9). In contrast, at a local scale the water cycle is an open system with a single input, precipitation (PPT), and two major outputs, namely **evapotranspiration (ET)** and **runoff**.

 KEY TERM

Runoff Here the term is used to describe all of the water leaving the land via river channel flow, overland flow and throughflow.

 KEY TERMS

Lithosphere The solid, rocky outer layer of the Earth, consisting of the crust, the outermost layer of the mantle and soil.

Evapotranspiration The term is used to describe a combination of evaporation and transpiration. Evaporation is the change in state of water from a liquid to a gas. For this to happen, heat energy is required. Evaporation can occur from the surface of any water store, including the ocean, surface water on the land and water intercepted temporarily on plant leaves. Many meteorological factors influence the rate of evaporation, including temperature, humidity and wind speed. Transpiration is the diffusion of water from vegetation into the atmosphere involving a change from liquid to gas. Water is lost through the stomata (pores) of leaves, and different types of vegetation vary greatly in terms of the rates of transpiration which are allowed.

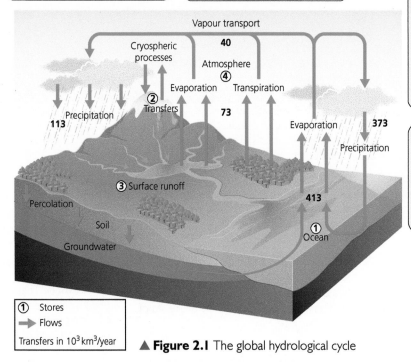

① In the oceans the vast majority of water is stored in liquid form, with only a minute fraction as icebergs.

② In the cryosphere water is largely found in a solid state, with some in liquid form as meltwater and lakes.

③ On land the water is stored in rivers, streams, lakes and groundwater in liquid form. It is often known as blue water, the visible part of the hydrological cycle. Water can also be stored in vegetation after interception or beneath the surface in the soil. Water stored in the soil and vegetation is often known as green water, the invisible part of the hydrological cycle.

④ In the atmosphere, water largely exists as vapour, with the carrying capacity directly linked to temperature. Clouds can contain minute droplets of liquid water or, at a high altitude, ice crystals, both of which are a precursor to rain.

Within the figure:

Vapour transport

Cryospheric processes

Atmosphere ④

40

Evaporation   Transpiration

② Transfers

73

Precipitation
113

Evaporation   373

Precipitation

③ Surface runoff

413

Percolation

Soil

① Ocean

Groundwater

① Stores
➡ Flows
Transfers in 10^3 km³/year

▲ **Figure 2.1** The global hydrological cycle

The water balance summarises the changes in the hydrological cycle during a particular time, and is expressed by:

$$P = Q + ET +/-\Delta\ SS +/-\Delta SMS +/-\Delta AZS +/-\Delta GS +/- DT$$

Where P is precipitation, Q is runoff, ET is evapotranspiration, Δ SS refers to changes in surface storage, ΔSMS refers to changes in soil moisture storage, ΔAZS refers to changes in aeration zone storage, ΔGS refers to groundwater storage and DT refers to the deep transfer of water across the watershed.

A simplified version of the formula is:

$$P = Q + ET +/- \Delta S$$

Where P = precipitation, Q = runoff, ET = evapotranspiration and ΔS = net positive or negative changes in total water storage.

Water can be stored at a number of places within the cycle. These stores include organisms (water forms part of biomass and may also be stored temporarily on vegetation surfaces after rainfall); depressions on the Earth's surface; soil moisture; groundwater and water bodies, such as rivers; and lakes. Water is stored in large amounts in bodies of ice such as glaciers and ice caps, known collectively as the cryosphere. The global hydrological cycle also includes stores in the oceans and the atmosphere.

Solar radiation drives the hydrological cycle. This is because the main source of energy available to the planet is the Sun. In some places there are important local sources of heat, for example geothermal heat in Iceland (see page 167) and human-related (anthropogenic) sources in large-scale urban-industrial zones. However, solar heating is the main driver of system operations for the hydrological cycle, as well as the main cause of global temperature patterns and global wind patterns.

The global hydrological cycle is complex. To enable our understanding of how the cycle works, scientists use a systems model. This is a simplified structuring of reality which shows just a few parts of the system (see page 4). Figure 2.3 shows a model of the hydrological cycle, although it may look quite unfamiliar, in comparison to the more typical presentation of the water cycle (Figure 2.1).

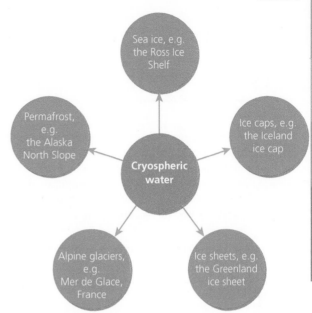

▲ **Figure 2.2** Water in the cryosphere

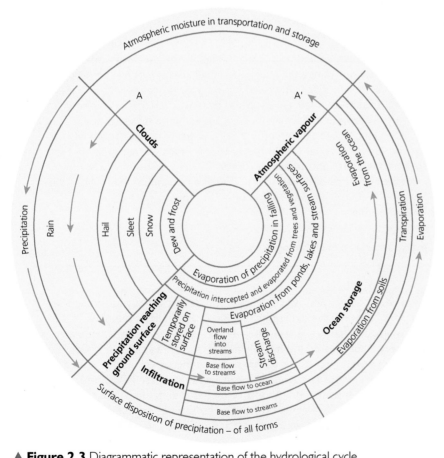

▲ **Figure 2.3** Diagrammatic representation of the hydrological cycle

ANALYSIS AND INTERPRETATION

(a) Describe how water is transferred through the hydrological cycle from point A to point A¹ in Figure 2.3.

GUIDANCE

It is important to use precise terminology in the description of movement in the hydrological cycle. For example: water at A falls as rain, and on reaching the surface infiltrates (soaks) into the soil. It then percolates deeper into the bedrock to become groundwater, and this makes its way to the streams (groundwater flow). The stream flow makes its way to the oceans where it is stored temporarily, until it is evaporated back into the atmosphere. There, condensation may occur and the water is returned to the ground in the form of precipitation.

(b) Suggest how hydrological cycle operations might differ between a mountainous region and a lowland rainforest region.

GUIDANCE

Although it is common to describe the hydrological cycle as a single system, there are many regional and seasonal variations in its operation. For example, the pattern of flows and stores in a tropical rainforest (see page 196) is very different from that of Arctic tundra (see page 204) or a high mountain region; the hydrological cycle at high latitudes and altitudes may have far more distinctive seasonal changes between winter and summer, and even between day and night.

In this question, it is very easy to over-generalise and so it would be helpful to give some indication of the height of the mountains involved. For example: in a high mountain area, there may be higher amounts of precipitation (orographic or relief rainfall) and/or more of it may fall as snow (which may remain on the ground until it thaws). Evaporation rates are likely to be low. There may also be significant diurnal variations between freezing and thawing. Much will depend on how high the mountain is – very high mountains, such as Mount Everest (over 8000 m high), remain frozen all year, but lower mountains, such as in the Alps or Rockies (c.3000–4000 m), may experience greater diurnal and seasonal variations in temperature, and so melting, freezing and evaporation. In contrast, a rainforest region is likely to receive high rates of rainfall (over 2000 mm per annum) throughout the year. The constant high temperatures lead to high evaporation rates, but the high density of forests reduces overland runoff and promotes infiltration.

(c) Explain two ways in which human activities alter the hydrological cycle.

GUIDANCE

There are many potential ways in which humans influence the hydrological cycle and you should try to provide a structured analysis. This could involve: by comparing direct and indirect methods; outlining short-term and long-term impacts; contrasting small-scale and large-scale effects. Human activities may have direct and indirect influences on the hydrological cycle, for example. Irrigation is a direct effect, in which humans extract water from stores (lakes, rivers, and groundwater) or trap it behind dams, and then divert the water towards farmland. This can lead to increased evaporation losses, salinisation in soils, reduced stores in rivers, etc. An example of an indirect impact is the conversion of land use from forest to farming. This generally leads to reduced interception, lower infiltration rates, reduced transpiration, increased overland runoff and reduced rainfall in the area.

Solar radiation provides the energy that drives the hydrological cycle. Although the amount of water in the system does not change, it can exist as a solid, liquid or gas, and some of these phase changes may last for millions of years. The basic processes involved in the cycle are evaporation, condensation and precipitation.

Changes in the state of water and the redistribution of water in the Earth/atmosphere system have implications for the Earth's energy budget. The evaporation of water requires the uptake of energy to cause the conversion from a liquid to a gas. As long as water remains in the form of water vapour, it retains that energy in the form of latent heat. The energy is released into the environment when the conversion back to a liquid takes place. Vapour in the atmosphere is easily moved, and the release of heat may take place thousands of kilometres from where it was absorbed. For example, during the summer monsoon in east Asia, a significant amount of water vapour is transferred from the southern oceans to continental Asia.

Evaporation and transpiration are easily disrupted by human activities. Since the cycle is a closed system with regard to water, human activities do not deplete the entire system, but can lead to the transfer of water between different parts of the system and can cause local water shortages.

Global water stores

Only a small fraction (2.5 per cent by volume) of the Earth's water supply is fresh water (Table 2.1). Of this, around 70 per cent is in the form of cryospheric storage (ice caps and glaciers), around 30 per cent is groundwater and the rest is made up of lakes, soil water, atmospheric water vapour, rivers and biota. Water on the surface of the Earth to which we have direct access (freshwater lakes and rivers) is only around 0.3 per cent of the total. Atmospheric water vapour contains only around 0.001 per cent of the Earth's total water volume. Taken together, all the forms in which the Earth's water can exist are called the hydrosphere.

Reservoir		Value (km^{-3} × 10^{-3})	% of total
Ocean		1,350,000.0	97.403
Atmosphere		13.0	0.00094
Land		35,977.8	2.596
Of which	Rivers	1.7	0.00012
	Freshwater lakes	100.0	0.0072
	Inland seas	105.0	0.0076
	Soil water	70.0	0.0051
	Groundwater	8200.0	0.592
	Cryosphere	27,500.0	1.984
	Biota	1.1	0.00088

▲ **Table 2.1** Global water reservoirs

KEY TERMS

Condensation The process by which vapour changes into a liquid or solid form. For this to happen in the atmosphere, condensation nuclei must also be present.

Precipitation Rainfall, sleet and snow are the products of cloud droplet growth. An average rain droplet is one million times larger than a cloud droplet. In order for rain to fall, cloud droplets have to undergo a rapid process of growth or fusion. The two recognised theories of rainfall production are (1) the Bergeron-Findeisen process of ice-crystal growth, and (2) the collision process, wherein 'super' condensation nuclei generate large heavy water droplets which collide with smaller droplets, sweeping them along into their wake.

Energy budget The state of balance between incoming solar radiation received by the atmosphere and the Earth, and the re-radiated heat or reflected energy.

Groundwater Water stored in solid rock and in any superficial deposits, e.g. gravels below the soil.

Hydrosphere Liquid or gaseous water found on, under and above the surface of a planet as groundwater, surface water (oceans, rivers and lakes) and atmospheric water vapour.

The different forms of water in the Earth's water budget are fully replenished (restored after use) during the hydrological cycle but at very different rates. The time for a water molecule to enter and leave a part of the system (i.e. the time taken for water to completely replace itself in part of the system) is called the turnover time.

Table 2.2 shows turnover times as crude averages – think carefully about why these average values might actually vary from place to place (dependent on climate and weather), or over time on account of natural climate changes (for instance, after the transition from the Pleistocene to the Holocene, turnover times would have changed greatly in Europe). Turnover times will also vary between drainage basins and river valleys according to differences in the size of particular rivers or the thickness and stability of any glacial ice masses.

Water location	Turnover time
Polar ice caps	10,000 years
Ice in the permafrost	10,000 years
Oceans	2500 years
Groundwater	1500 years
Mountain glaciers	1500 years
Large lakes	17 years
Peat bogs	5 years
Upper soil moisture	1 year
Atmospheric moisture	12 days
Rivers	16 days
Biological water	A few hours

▲ **Table 2.2** Turnover time for different parts of the hydrosphere

Human impact on the global hydrological cycle

Of the 8 per cent of the annual freshwater runoff that is withdrawn for human use, over 70 per cent is returned to the water cycle through evaporation and the remainder is eventually returned as partially degraded water. Of the annual freshwater runoff (some 40,000 km^3), over 38,000 km^3 makes it to the oceans, either directly (92 per cent) or indirectly, as partially degraded (polluted) water.

The greatest human use is for irrigation, some 7 per cent of the annual freshwater runoff (Figure 2.4). Of this, over two-thirds is evaporated to the atmosphere and, after use, the rest runs off as partly degraded water. Industrial use is the next largest use (but only less than 10 per cent of the amount used for irrigation). A smaller percentage is evaporated than leaves through runoff. Finally, domestic and commercial land uses account for the least volume of withdrawals, and of this water, an equal amount of water is evaporated and lost via runoff.

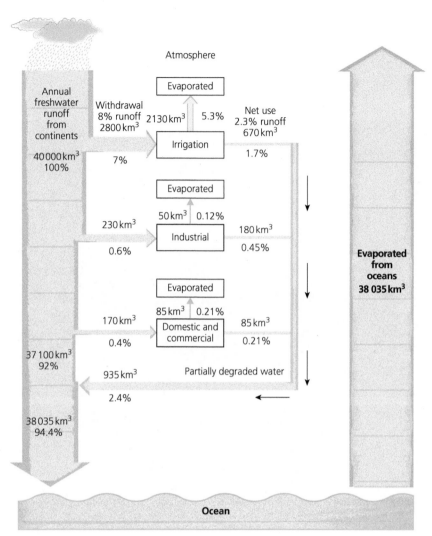

▲ **Figure 2.4** The effects of humans on the global water cycle

The degree to which water can be seen as a renewable or non-renewable resource depends on where it is found in the hydrological cycle. Renewable water resources are stores which are replenished yearly or more frequently in the Earth's water turnover processes. Thus, groundwater is a non-renewable source of water as turnover time is very long. An **aquifer** is an underground formation of permeable rock or loose material which stores groundwater. Aquifers can produce useful quantities of water when tapped by wells. Aquifers come in all sizes, from small (a few hectares in area) to very large (covering thousands of square kilometres). They may be only a few metres thick, or they may measure hundreds of metres from top to bottom. Intensive use of aquifers unavoidably results in depleting the storage and has unfavourable consequences: it depletes the natural resource and disturbs the natural steady state equilibrium established over centuries. Restoration requires tens to hundreds of years.

 KEY TERM

Aquifer A permeable rock layer which contains water that can be extracted for human use.

▲ **Figure 2.5 a** Large scale irrigation at Keiskammahoek, South Africa **b** Irrigation and water storage in the Eastern Cape, South Africa

CONTEMPORARY CASE STUDY: NEW FINDINGS ABOUT THE GLOBAL HYDROLOGICAL CYCLE

In 2017, a team of German scientists announced their finding of a reservoir of water deep beneath the Earth's surface which represents three times the volume of all water in the oceans. The water is contained within a blue rock called ringwoodite that lies 700 kilometres underground in the mantle. This discovery supports the idea that the oceans gradually oozed out from the interior of the Earth. It could be that the water that makes our planet habitable was present in the dust that coalesced to create Earth, rather than arriving later on ice-rich comets or asteroids.

Scientists used 2000 seismometers to study the seismic waves generated by more than 500 earthquakes to identify the water within the Earth's interior. The water layer revealed itself because the seismic waves slowed down, as it takes them longer to get through wet rock than dry rock.

The team of scientists found signs of wet ringwoodite in the transition zone that divides the upper and lower regions of the mantle. At that depth (700 km), the pressures and temperatures are sufficient to squeeze the water out of the ringwoodite. These findings support a recent study of a diamond from the transition zone that had been carried to the surface in a volcano, which also found that it contained water-bearing ringwoodite.

This water is much deeper than any seen before, at a third of the way to the edge of Earth's core. One conclusion that the scientists reached was that the water cycle on Earth is more complex than had previously been thought, extending into the deep mantle.

The Earth's interior may be dotted with similar wet pockets lurking below other major volcanoes. Scientists from Bristol have studied a huge 'anomaly' 15 kilometres beneath the dormant Uturuncu Volcano in the Bolivian Andes. The anomaly, called the Altiplano-Puna magma body, slows down seismic waves, unlike surrounding magma. They found that, at a particular water content, the electrical conductivity exactly matched the value measured in the anomaly. They calculated it contained 8 to 10 per cent water. The Altiplano-Puna magma body is known to be around half a million cubic kilometres in volume, so the researchers estimate it must contain a similar amount of water to some of the largest freshwater lakes on Earth.

Other anomalies with similar unexplained conductivity have been discovered beneath other volcanoes, such as those in the Taupo Volcanic Zone in New Zealand, and Mount St Helens in the USA.

However, this water cannot be used for human activities. It is dissolved in partially melted rock at 950 to 1000°C, so it is not accessible.

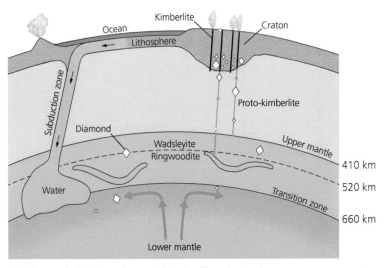

▲ **Figure 2.6** Water deep within the Earth's interior

② Local drainage basin dynamics

▶ *What are the main stores, system movements and water cycling processes in local drainage basins?*

The global hydrological cycle is very complex even when it is described using a systems approach. Many parts of the cycle cannot be observed. Hence scientists study the drainage basin as a subsystem (see page 2) of the hydrological cycle. It is visualised in Figure 2.7 as a cascading system (see page 4). The drainage basin is taken as the unit of study rather than the global system, in part because it is easier to observe and measure flow and store volumes reliably.

Drainage basin outputs and stores

KEY TERM

Leakage The transmission of groundwater at great depth from one drainage basin or aquifer towards another.

As previously discussed, the drainage basin water cycle has a single input, precipitation (PPT), and two major losses (outputs): evapotranspiration (ET) and runoff. Some hydrological models additionally identify a third output, leakage, that can occur from deep groundwater to other basins, especially if the rocks are faulted. The drainage basin system is an open system as it allows the movement of energy and matter across its boundaries. Table 2.3 shows varying drainage basin outputs.

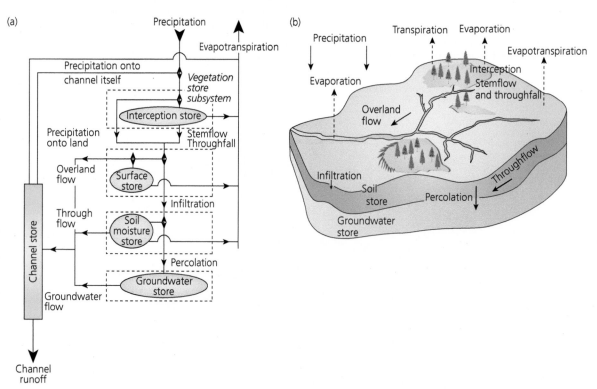

▲ **Figure 2.7** The drainage basin hydrological cycle visualised **a** as a cascading system and **b** in diagrammatic form

Water can also be stored at a number of stages or levels within the cycle. These stores include vegetation (both within and on the surface of), surface (depression storage), soil moisture, groundwater and water channels themselves. Table 2.4 shows drainage basin stores.

Human modifications to output and store patterns can always be observed at every geographical scale. Good examples include large-scale changes of

How it works	Spatial and temporal variations
Evaporation Evaporation is the change in state of water from a liquid to a gas. For this to happen, heat energy is required. • Evaporation can occur from the surface of any water store, including the ocean, surface water on the land and water intercepted temporarily on plant leaves. Globally, it is most important from oceans and seas.	• Evaporation increases under warm, dry, windy conditions and decreases under cold, calm conditions. Evaporation losses are greater in arid and semi-arid climates than in polar regions. • Factors affecting evaporation include meteorological factors such as temperature, humidity and wind speed. Of these, temperature is the most important factor. Other factors include the amount of water available, vegetation cover and colour of the surface (albedo or reflectivity of the surface).
Evapotranspiration (ET) The term is used to describe a combination of evaporation and transpiration. • Transpiration is the diffusion of water from vegetation into the atmosphere involving a change from liquid to gas. • Water is lost through the stomata (pores) of leaves, and different types of vegetation vary greatly in terms of the rates of transpiration which are allowed.	• ET represents the most important aspect of water loss, accounting for the loss of nearly 100 per cent of the annual precipitation in arid areas and 75 per cent in humid areas. Only over ice and snow fields, bare rock slopes, desert areas, water surfaces and bare soil will purely evaporative losses occur. • Transpiration varies with vegetation. Broad leaves have relatively high rates of transpiration whereas needle-shaped leaves and spines on cacti have evolved to minimise water loss. Transpiration also varies with crop type – grass-like crops such as corn and wheat transpire less than broad-leaved crops such as cabbages or turnips.
Potential evapotranspiration (P.ET) The distinction between actual ET and P.ET lies in the concept of moisture availability: • Potential evapotranspiration is the water loss that would occur if there was an unlimited supply of water in the soil for use by the vegetation.	• For example, the actual evapotranspiration rate in Egypt is less than 250 mm, because there is less than 250 mm of rain annually. However, given the high temperatures experienced in Egypt, if the rainfall reached 2000 mm, there would be sufficient heat to evaporate that amount of water! Hence the potential evapotranspiration rate there is 2000 mm. • The factors affecting evapotranspiration include all those that affect evaporation. In addition, some plants, such as cacti, have adaptations to help them reduce moisture loss.
River discharge This is the movement of water in channels such as streams and rivers. The water may enter the river as direct channel precipitation (it falls on the channel) or it may reach the channel by surface runoff, groundwater flow (base flow) or throughflow (water flowing through the soil).	• River discharge can vary seasonally – some monsoonal rivers experience peak discharge in summer whereas some high latitude and high altitude rivers have their peak following springmelt. • Other rivers have a maximum flow during winter, and others still have complex regimes which are a mix of different regimes (see page 52).

▲ **Table 2.3** Varying outputs (in space and time) from a drainage basin

channel flow (abstraction for human uses, discharge by industry/ wastewater treatment, changes in temperature, changes in ecology) and storage (dams), irrigation and land drainage, and large-scale abstraction of groundwater and surface water for domestic and industrial use.

How it works	Spatial and temporal variations
Interception Water that is caught and stored by vegetation has been intercepted. After interception, one of three things happens to the water: ● Interception loss – water that is retained by plant surfaces and which is later evaporated away or absorbed by the plant. ● Throughfall – water that either falls through gaps in the vegetation or which drops from leaves or twigs. ● Stemflow – water that trickles along twigs and branches and finally down the main trunk.	Interception loss varies with different types of vegetation (Figure 2.8). ● Interception is less from grasses than from deciduous woodland owing to the smaller surface area of the grass shoots. From agricultural crops, and from cereals in particular, interception increases with crop density. ● Coniferous trees intercept more than deciduous trees in winter, but this is reversed in summer. ● Marked seasonal changes in storage occur in **biomes** and ecosystems that are characterised by deciduous vegetation.
Soil water Soil water (soil moisture) exists in the subsurface soil layers above the water table. From here water may be absorbed, held, transmitted downwards towards the water table or transmitted upwards towards the soil surface and the atmosphere (for example, by a process called capillary action).	● In coarse-textured soils (found in granite uplands of the UK), much of the water is held in fairly large pores at fairly low suctions, while very little is held in small pores. ● In the finer-textured clay soils (typical of lowland floodplains), the range of pore sizes is much greater and, in particular, there is a higher proportion of small pores in which the water is held at very high suctions.
Field capacity and wilting point Field capacity is the amount of water held in the soil after excess water drains away after a rainfall event, i.e. saturation or near saturation. Whereas the term wilting point refers to the threshold level of soil water below which plants cannot extract sufficient water to maintain their mechanical strength (see Figure 2.9).	● The field capacity of a soil depends largely on its texture (Figure 2.9). Sandy soils have a low field capacity whereas clay soils have a high field capacity. Sandy soils also reach their field capacity very quickly due to the ease with which water can infiltrate into the soil and the low quantity of water required. ● In contrast, clay and **loam** soils take longer to reach field capacity due to the slow rate of infiltration and the large volume of water required to be absorbed to reach field capacity.

KEY TERMS

Biome A collection of ecosystems sharing similar climatic conditions – for example, tundra, tropical rainforest and desert.

Water table The upper surface of the zone of saturation in permeable rocks – it varies seasonally with the amount of percolation.

Loam A soil with a relatively even inorganic mix of sand, silt and clay.

How it works	Spatial and temporal variations
Aeration zone storage This term refers to the moisture in the soil that is immediately above the groundwater store. It is unsaturated and contains air and moisture in its pore spaces.	This zone varies seasonally and depends on the depth of the water table. Following heavy rains and springmelt, there should be an increase in the amount of water percolating down towards the water table. Human withdrawal of groundwater may make the aeration zone deeper.
Surface water There are a number of types of surface water, some of which are temporary and some are permanent.	• Temporary sources include small puddles following a rainstorm and turloughs (seasonal lakes in limestone in the west of Ireland that develop during the winter due to higher water tables). • Permanent stores include lakes, wetlands, swamps, peat bogs and marshes. • Surface water storage is impacted by the building of dams and the creation of reservoirs.
Groundwater Groundwater refers to subsurface water that is stored under the surface in rocks. • Groundwater accounts for 22.7% of all freshwater on the Earth (Table 2.1). However, while some soil moisture may be recycled by evaporation into atmospheric moisture within a matter of days or weeks, groundwater may not be recycled for as long as 20,000 years. • The term groundwater recharge refers to the refilling of water in pores where the water has dried up or been extracted by human activity.	• In small drainage basins, seasonal variations in groundwater may be more noticeable than in large drainage basins. • Due to extraction by human activities many of the world's major groundwater stores are becoming depleted. The Ogallala aquifer of the High Plains of Texas has experienced a fall in the water table of 30–50m in less than 50 years. The aquifer has narrowed by more than 50%, and there have been reductions in the area irrigated and in crop yields. • The aquifer under Saudi Arabia is a fossil aquifer in that it is not being replenished. The aquifer is predicted to be dry by 2030.
Channel storage Channel storage refers to all water that is stored in rivers, streams and other drainage channels.	• Some rivers are seasonal, while others may disappear underground either naturally, e.g. in areas of carboniferous limestone, or in urban areas, where they may be covered (culverted – for example, the 'lost' rivers of London such as the Fleet and the Effra).

▲ **Table 2.4** Stores in a drainage basin

🔑 **KEY TERM**

Groundwater recharge
The movement (percolation) of water from the surface into an aquifer.

ANALYSIS AND INTERPRETATION

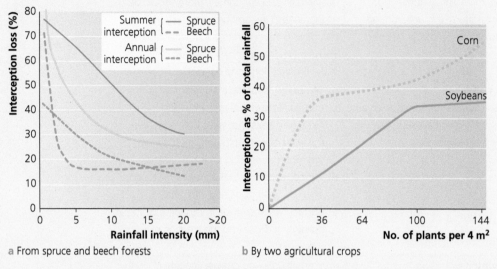

a From spruce and beech forests   b By two agricultural crops

Source: *Advanced Geography: Concepts & Cases* by P. Guinness & G. Nagle (Hodder Education, 1999), p.245

▲ **Figure 2.8** Interception losses for different types of vegetation

Study Figure 2.8, which shows interception losses linked with varying amounts of rainfall for several different types of vegetation and crop cover.

(a) Analyse the seasonal variation in interception between spruce trees and beech trees.

GUIDANCE

A good answer to this question will adopt an analytical style which identifies clearly the 'big story' these data are showing, along with supporting evidence. For example, spruce intercepts over 70 per cent of limited rainfall (< 5 mm) throughout the year, falling to about 30 per cent of rainfall when rainfall reaches round 20 mm. In contrast, during the summer, spruce intercepts between 50 per cent and 80 per cent of low rainfall, and about 25 per cent of higher rainfall. Beech intercepts over 30 per cent of annual rainfall of less than 5 mm, but only about 15 per cent of higher rainfall. However, it intercepts nearly 70 per cent of rainfall of less than 1 mm, falling to less than 20 per cent of rainfall of over 5 mm.

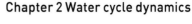

(b) Suggest reasons for these variations.

(c) Explain why the interception rate of corn and soybeans have non-linear curves in Figure 2.8b.

Drainage basin flows

Above-ground system transfers

Following an input of precipitation to the drainage basin, water may begin flowing from the interception store to the ground surface below:

- Throughfall refers to water which either falls through gaps in the vegetation or which drops from leaves or twigs.
- Stemflow refers to water which trickles along twigs and branches and finally down the main trunk.

Next, overland flow (also called surface runoff) may occur if environmental factors and conditions favour its operation. Overland flow can occur after either very intense or high duration rainfall:

1 **Saturation-excess overland flow** happens if rainfall continues for a long time.

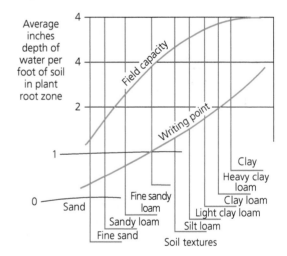

▲ **Figure 2.9** Field capacity and the wilting point according to soil type

KEY TERM

Overland flow (surface runoff) A visible movement of a sheet of water over the ground surface. Overland flow may be produced in many ways, including an excess of rainfall over infiltration capacity and an excess of rainfall over soil storage capacity. Overland flow is common in areas of high precipitation intensity and low infiltration capacity.

All soil layers become saturated and throughflow is deflected closer and closer to the surface. This is because over time during a rainfall event, the upper and more permeable soil horizons become saturated as the water level in the soil rises (exactly the same thing happens if you pour too much water into a plant pot: the level of standing water rises until it reaches the surface). In time the entire soil becomes saturated right up to the surface. Thereafter, saturation overland flow begins.

2   **Infiltration-excess overland flow** is defined as overland flow which occurs when rainfall intensity is so great that not all water can infiltrate, *irrespective of how dry or wet the soil was prior to the rainfall event*. This process, first described by R.E. Horton in 1945, is extremely common in some parts of the world, especially semi-arid areas where high-intensity rainfall encounters hard-baked ground with a relatively low infiltration capacity. In these environments, infiltration-excess overland flow can lead to flash flooding and the growth of large, deep channels called *wadis* which fill with water when it rains. In contrast, the process is less common in parts of the world with a humid temperate climate, such as the UK. You may, however, have witnessed a rare infiltration-excess overland flow event following a torrential downpour of summer rain which falls so fast that a sheet of water begins immediately to flow across the surface of the land: there is simply no time for the water to soak into the ground or run into drains.

Figure 2.10 shows the operation of both types of overland flow.

In areas of high precipitation intensity and low infiltration capacity, overland runoff is common. This is clearly seen in semi-arid areas and in cultivated fields. By contrast, where precipitation intensity is low and infiltration is high, most overland flow occurs close to streams and river channels. As a storm progresses, the area that produces overland flow extends up-valley from the stream. This is known as the partial area contribution model.

Overland flow ultimately contributes to channel flow (also called stream flow), i.e. the movement of water in channels such as streams and rivers. The moving water occupying the channel may have entered the stream as a result of direct precipitation, overland flow, groundwater flow or throughflow (see page 44).

Below-ground system transfers

Water is transmitted into the soil under the influence of gravity and, if local conditions favour further downwards movement, into the bedrock below.

Infiltration into the soil moisture store

Infiltration is the process by which water soaks into or is absorbed by the soil.

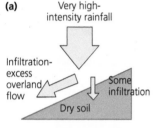

(a) Very high-intensity rainfall

Infiltration-excess overland flow

Some infiltration

Dry soil

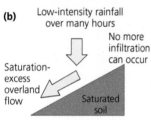

(b) Low-intensity rainfall over many hours

No more infiltration can occur

Saturation-excess overland flow

Saturated soil

▲ **Figure 2.10** The differences between infiltration-excess and saturation overland flow

🔑 **KEY TERM**

Infiltration The movement of water vertically downwards into the soil. The rate at which water enters the soil (the infiltration rate) depends on the intensity of rainfall, the permeability of the soil and the extent to which it is already saturated with water.

- The infiltration capacity of a soil is the maximum rate at which rain can be absorbed in a given condition (such as whether the soil is dry or already close to saturation).
- Infiltration capacity decreases with time through a period of rainfall until a more or less constant value is reached (Figure 2.11). Infiltration rates of 0–4 mm/hour are common on clays whereas 3–12 mm/hour are common on sands. Vegetation also increases infiltration. This is because it intercepts some rainfall and slows down the speed at which it arrives at the surface. For example, on bare soils where rainsplash impact occurs, infiltration rates may reach 10 mm/hour. On similar soil types covered by vegetation, rates of between 50 and 100 mm/hour have been recorded. Infiltrated water is chemically rich as it picks up minerals and organic acids from vegetation and soil.

The infiltration capacity of soils is usually great at the start of a rainfall event which has been preceded by dry conditions, but decreases rapidly as the rain continues to fall and water soaks into the soil. After several hours, the infiltration rate has become almost constant. The reason for the high starting value, and its rapid decline, is that the soil pores have become filled with water or with particles carried down from above. In addition, clay particles swell when hydrated and so pore spaces in clay soils decline in size, thereby reducing infiltration. Thus, sandy soils experience little decline in infiltration rate whereas clay-rich soils quickly become saturated and infiltration rates decline markedly. This is shown in Figure 2.13a.

Moreover, some sandy soils can maintain a high infiltration capacity after prolonged, heavy rain without generating any overland flow, unlike clay-rich soils (Table 2.5). Many forms of artificial disturbance of soils, such as agriculture and land-use changes, will decrease the infiltration capacity and promote overland flow (Figures 2.13b and 2.13c).

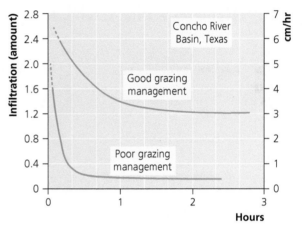

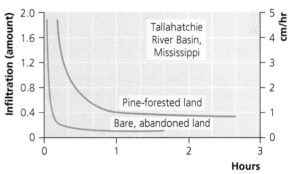

▲ **Figure 2.11** Infiltration rates vary greatly according to soil texture and land use

- Cultivation exposes the soil and makes it vulnerable to raindrop impact, thereby sealing the surface and promoting overland runoff.
- Similarly, fire may destroy the vegetation cover and increase raindrop impact.
- Trampling by livestock compacts the soil into an impermeable layer and promotes overland runoff.

Ground cover	Infiltration rate (mm/hour)
Old permanent pasture	57
Permanent pasture: moderately grazed	19
Permanent pasture: heavily grazed	13
Strip-cropped	10
Weeds or grain	9
Clean tilled	7
Bare, crusted ground	6

▲ **Table 2.5** Influence of ground cover on infiltration rates

ANALYSIS AND EXPLANATION

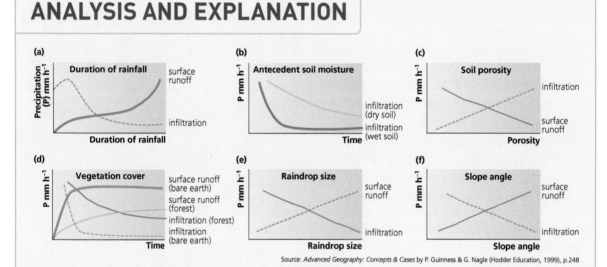

Source: *Advanced Geography: Concepts & Cases* by P. Guinness & G. Nagle (Hodder Education, 1999), p.248

▲ **Figure 2.12** Factors affecting infiltration and surface runoff

Figure 2.12 shows how infiltration is inversely related to overland runoff. The relationship is influenced by a variety of factors: duration of rainfall, antecedent soil moisture (pre-existing levels of soil moisture), soil porosity, vegetation cover, raindrop size and slope angle.

Briefly describe and suggest reasons for variations in infiltration with (a) duration of rainfall, (b) antecedent soil moisture, (c) soil porosity, (d) vegetation cover, (e) raindrop size and (f) slope angle.

(a) Duration of rainfall

GUIDANCE

As the length of the rainfall event increases, the rate of infiltration decreases and surface runoff increases. This is because, as rainfall progresses, there is an increasing amount of water in the soil, therefore reducing the potential for infiltration. If less water can enter the soil, more flows over the surface.

(b) Antecedent soil moisture

GUIDANCE

If the soil is relatively dry, there is more potential for infiltration to occur. However, initially there is little difference between infiltration in dry and wet soils. This may be because there are still pore spaces near the surface which contain air as the water already in the soil has moved downwards. Once these pore spaces are filled with water, infiltration into the wet soil decreases, but the dry soil still has the potential to store more water.

(c) Soil porosity

GUIDANCE

As soil porosity increases, infiltration increases and surface runoff decreases. There is a positive relationship between porosity and infiltration and an inverse relationship between infiltration and surface runoff. This is due to the large size of the pore spaces – large pore spaces allow more water to infiltrate than small pore spaces.

(d) Vegetation cover

GUIDANCE

Vegetation has a major impact on infiltration and surface runoff. Surface runoff is rapid over bare earth and is lower under forest cover. In contrast, it is high over bare earth initially, but declines rapidly. This is most likely because the pore spaces in the upper layers of the soil have filled up with water. However, under vegetated ground (forest) infiltration decreases more slowly – largely due to the lower amount of water getting through the trees and leaves (interception) and the reduced speed with which the water reaches the ground.

(e) Raindrop size

GUIDANCE

As raindrop size increases, infiltration decreases – this may be due to the size of the raindrop (large in high-intensity rainfall events – convectional storms) exceeding the size of the pore spaces they are entering (the infiltration capacity). As infiltration decreases, surface runoff increases.

(f) Slope angle

GUIDANCE

As slope angle decreases, the influence of gradient increases, infiltration rate decreases and overland runoff increases. On low angle slopes, the rate of infiltration is higher but rates of overland flow are lower. This is because gradient affects the amount of time that water remains on a slope – on low angle slopes, water remains for longer, and so there is more potential for infiltration.

KEY TERMS

Throughflow This is the movement of water laterally (sideways) through the soil, via a matrix of pore spaces, fissures and pipes (wide gaps within the soil created by roots and animal burrows). With the exception of pipeflow, throughflow movements are relatively slow. Throughflow is most effective in the surface horizons of the soil because these are, in general, less compacted and have high permeability.

Base flow The movement of water from land to river through saturated rock, i.e. below the water table. It is the slowest form of such water movement, and accounts for the constant flow of water in rivers during times of low rainfall. This is the reliable background river flow of a drainage basin, and in some basins may be aided by interflow from above the water table and throughflow through the soil.

Percolation The vertical movement of water down through the soil or the bedrock in the unsaturated (vadose) zone.

Groundwater flow The vertical and lateral movement of water through a drainage basin's underlying rock as a result of gravity and pressure.

Throughflow and interflow

Throughflow is the movement of water laterally (sideways) through the soil store, via a matrix of pore spaces, fissures (called percolines) and pipes (wide gaps within the soil created by roots and animal burrows).

- With the exception of pipeflow, throughflow movements are relatively slow. Throughflow is most effective in the surface horizons of the soil because these are, in general, less compacted and have high permeability.
- On farmed land, the surface horizon has often been tilled (ploughed) and so the soil structure is open, with many large spaces and fissures which water can soak into.
- In contrast, lower soil horizons are more compacted by the weight of overlying material, which reduces the soil's permeability. When this happens, water which is soaking downwards under gravity is deflected laterally down slope within the soil.

Interflow is a secondary, deeper flow of water through the soil. Interflow responds to rainfall more slowly than surface runoff but more rapidly than **base flow**. Throughflow and interflow movements often feed directly into the river channel store as Figure 2.7 (page 34) indicates.

Percolation and groundwater movements

Water may move slowly downwards from the soil into the bedrock below – this is known as **percolation**. Depending on the permeability of the rock, this may be very slow, or in some rocks, such as carboniferous limestone and chalk, it may occur quite quickly. As it fills, the groundwater store then begins to generate water movements of its own. This is called **groundwater flow**; it is an important water movement because it maintains the base flow of rivers during periods of low rainfall.

All rocks show some porosity (the volume of voids as a percentage of the bulk volume of material) and resulting permeability. However, porosity and permeability vary enormously according to rock type. Coarse-grained sedimentary rocks show high permeability and permit high levels of groundwater flow, whereas fine-grained igneous rocks are relatively impermeable. Below a certain depth, the bedrock of a drainage basin may be permanently saturated. This level below which the ground is saturated is called the surface of the water table and it can vary in depth according to the season.

Groundwater and aquifer characteristics

Most groundwater is found within a few hundred metres of the surface but has been found at depths of up to 4 kilometres beneath the surface. The permanently saturated zone within solid rocks and sediments is known as the phreatic zone. The upper layer of this is known as the water table. The water table varies seasonally. It is higher in winter following increased levels of precipitation. The zone that is seasonally wetted and seasonally dries out is known as the aeration zone. Some groundwater discharge areas may be perennial (flowing year after year).

Aquifers, as we have seen (Table 2.1, page 29), are significant underground reservoirs of water.

- The water in aquifers moves very slowly and acts as a natural regulator in the hydrological cycle by absorbing rainfall which otherwise would reach streams rapidly.
- Aquitards are rocks of low permeability which slow down water movement (Figure 2.14).
- In contrast, aquicludes are impermeable rocks which prevent the downward movement of water.

Aquifers help maintain stream flow during long dry periods. Where water flow reaches the surface (as shown by the discharge areas in Figure 2.13), springs are sometimes found. These may be substantial enough to become the source of a stream or river.

Groundwater recharge of aquifers occurs naturally over varying timescales as a result of:

- infiltration of part of the total precipitation at the ground surface
- seepage through the banks and bed of surface water bodies such as ditches, rivers, lakes and oceans
- groundwater leakage and inflow from adjacent rocks and aquifers
- artificial recharge methods and percolation from water used for irrigation.

Losses of groundwater can result from high rates of evapotranspiration, particularly in low-lying areas where the water table is close to the ground surface; it also occurs due to natural discharge by means of spring flow and seepage into surface water bodies. Groundwater may also leak and flow out through aquicludes and into adjacent aquifers, and it may be lost through artificial abstraction. For example, the water table near Lubbock on the High Plains of Texas (USA) has declined by 30–50 m in just 50 years, and in Saudi Arabia the groundwater reserve in 2010 was 42 per cent less than in 1985. Aquifers are highly dynamic water stores whose functioning can be adversely affected by all of the physical and human factors operating on varying temporal and spatial scales.

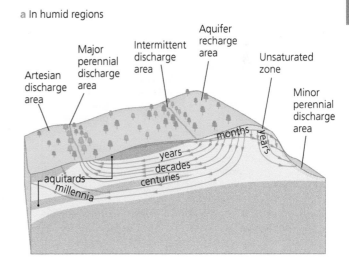

a In humid regions

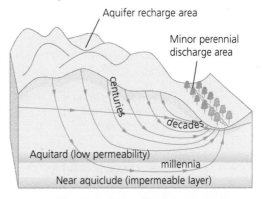

b In semi-arid regions

Source: *Advanced Geography: Concepts & Cases* by P. Guinness & G. Nagle (Hodder Education, 1999), p.248

▲ **Figure 2.13** Groundwater and aquifer characteristics

③ Urbanisation and the hydrological cycle

▶ *What are the impacts of urbanisation on the hydrological cycle?*

KEY TERM

Urbanisation An increase in the percentage of a population living in towns and cities, usually resulting in an increase in the urban footprint size and thus the extension of impermeable artificial surfaces.

Urbanisation has a significant effect on the functioning of the hydrological cycle. It interrupts and rearranges the inputs, storage and transfer of water. Moreover, there is a clear link between urbanisation and key problems of declining water resources, increasing pollution and increased risk of flooding.

The hydrology of urbanisation

Geographers are increasingly aware of the effect of urbanisation on hydrology (Figure 2.14). In urban areas, vegetated soils are replaced by impermeable surfaces. These typically cover 20 per cent or more of post-war urban areas. In some central areas, such as city centres, it can be as high as 90 per cent. By contrast, in areas of suburban detached housing, it can be as low as 5 per cent. This change can lead to a number of effects, including: reduced water storage on the surface and in the soil; increased percentage of catchment runoff; increased velocity of overland flow; decreased evapotranspiration (since urban surfaces are usually non-vegetated and dry); and reduced percolation to groundwater because the surface is impermeable.

Additionally, urbanisation brings major changes in the drainage density of an area (drainage density refers to the total length of stream channel per square kilometre). The channel network is increased by stormwater sewers, gutters, gullies and drains (Figure 2.15). Prior to urbanisation the stream channel network would have been much more limited. The increase in drainage density has a number of effects, including reduced distance of overland flow; increased velocity of flow because sewers are smoother than natural channels; and reduced storage in the channel system, as sewers are designed to drain as completely and as quickly as possible.

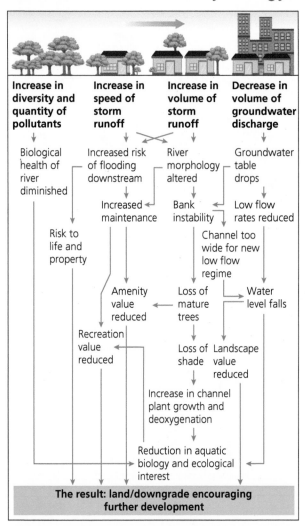

▲ **Figure 2.14** The effects of urbanisation on hydrological systems and the wider physical environment

▲ **Figure 2.15** Impermeable conditions and increased drains combine to increase the risk of flooding in urban areas

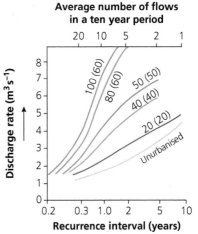

Changes in the magnitude and frequency of flood discharges during urbanisation

Numbers indicate percentage of area:
• sewered
• impervious (in brackets)

▲ **Figure 2.16** Urbanisation and flood frequency curves

In addition, there are rapid increases in rates of erosion during periods of construction. During the building of houses, roads and bridges, vegetation is cleared. This exposes the soil to storms and allows increased amounts of overland flow. Heavy machinery disturbs and churns the soil, which increases its erodibility. However, some activities may bury the soil under concrete, tar or tiles. This effectively stops any further erosion of the soil.

ANALYSIS AND INTERPRETATION

(a) Study Figure 2.16. Outline the relationship between flood recurrence and the average number of floods over a ten-year period in an unurbanised catchment, making reference in your answer to the scale used in the graph.

GUIDANCE

Firstly, identify the unurbanised (yellow) line! As you can see, the flood recurrence interval and the average number of floods over a ten-year period are positively correlated. For a flood recurrence interval of ten years (i.e. one flood every ten years), the average number of floods will be just one. If the flood recurrence interval is every five years, on average there should be two floods every ten years. In contrast, if the flood recurrence is every 0.5 years, we would expect two floods per year, or twenty floods over a ten-year period. The scale used is logarithmic as it includes some very low values (0.2, 0.5) as well as some relatively high ones (e.g. 10).

(b) Explain, using data, why flood risk alters as changes occur in the percentage of an area that is impervious and uses sewer pipes.

GUIDANCE

The impervious (impermeable) areas in an urban area do not absorb water, rather they repel it. This means it either sits on the surface or is taken away. The more drainage channels there are (another way of interpreting the addition of storm sewers, drains and gutters), the more likely it is that water will be carried away rapidly into a river channel or storm relief channel. Thus, for an area that is 20 per cent sewered, the discharge is likely to be low – about 2 m/s for a one-year flood, and about 5 m/s for a five-year flood. In contrast, for an urban area that is completely sewered, the one-year flood is likely to be around 7–8 m/s and the five-year flood over 9 m/s. Thus, the size of the flood varies with the percentage of the area that is impervious and the percentage of the area that is sewered.

(c) Evaluate the information that this graph provides.

GUIDANCE

The command word 'evaluate' requires you look at some strengths and weaknesses before, ideally, providing a single-sentence summing-up. For example:

■ The data that the graph provides are very insightful because they show how different urban characteristics produce flood risks of varying sizes. Some urban areas are more built up than others (in terms of impermeable areas and percentage sewered). We might expect there to be a difference between high-income countries and low-income countries, but even within either category, there are important differences between urban areas in countries that have lots of space and suburban sprawl, e.g. Australia and USA, and those that are more compact, e.g. Japan. Indeed, even within the same city, especially megacities, there may be significant differences in the level of impermeability between city centres and the suburbs.

■ However, the graph suggests that all the differences are due to two factors. This is a simplification. Other factors are also important, e.g. the size of the storm, its length, total amount of rainfall, intensity of rain and vegetation cover, among others, all play a part. The graph is also quite complex and not necessarily easy to understand for those without some scientific training.

Overall, the graph is useful, as it suggests that not all urban areas are the same, and that even within a single urban area, there is considerable variation in the flood risk.

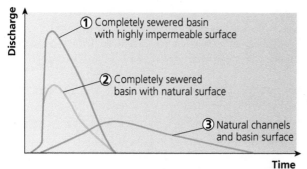

Source: *Advanced Geography: Concepts & Cases* by P. Guinness & G. Nagle (Hodder Education, 1999), p.255

▲ **Figure 2.17** The effects of urban development on storm hydrographs

On all urban impermeable surfaces there is initial wetting of the surface, perhaps some absorption of water by the surfaces, and certainly the filling of surface depressions with water (sometimes called pluvial flooding). After virtually all rainstorms, there will be evaporation from the surface into the atmosphere. Roofs are free from the infiltration of water but when the capacity of their gutters is exceeded in a severe storm there is overland flow to the ground.

Supporting evidence for this comes from a study which showed a 530 per cent increase in the paved area of Long Island, New York, led to a 270 per cent increase in total runoff. Another study demonstrated annual runoff from the city centre of Kursk in Russia is 109 mm while just 45 mm of water drains off in this way from suburban catchments. In the surrounding countryside, runoff varies between 7 mm and 14 mm, depending on the cropping regime, while in natural grasslands the runoff falls to zero.

So is it the case that urbanisation *always* increases water flow rates in drainage basins? Actually, it can also be argued that *after continued heavy rainfall* there is no hydrological difference between tar and saturated soil. Beyond a certain threshold, which is hard to determine, the land use has little effect on flood magnitude.

River hydrographs and river regimes

▶ *Why do river hydrographs and regimes vary from place to place and time to time?*

A **storm hydrograph** shows how the discharge of a river varies over a short time (Figure 2.18). Normally it refers to an individual storm or group of storms of not more than a few days in length. Before the storm starts the main supply of water to the stream is through groundwater flow or base flow. This is the main supplier of water to rivers. During the storm some water infiltrates into the soil while some moves quickly over the surface as overland flow, causing a rapid rise in the level of the river. The rising limb shows us how quickly the flood waters begin to rise whereas the recessional limb is the speed with which the water level in the river declines after the peak. The **peak discharge (or peak flow)** is the maximum discharge of the river as a result of the storm and the **lag time** is the time between maximum rainfall and the maximum flow in the river.

The characteristics of storm hydrographs vary with a number of key variables (Table 2.6). For example, as gradient increases, the peak flow increases and the lag time decreases. This is because an increased gradient leads to a greater amount of overland flow and faster rates of overland flow. Similarly, as the number of streams or channels in an area increases (i.e. the drainage density), peak flow is increased and lag time is decreased. This is because more water gets into streams (because there are more of them) and the channels transport the water rapidly away into the main river.

 KEY TERMS

Storm hydrograph A graph showing how the discharge of water in a river varies over time following a storm event or events.

Peak discharge (or peak flow) The greatest rate of flow reached by the stream during a storm event.

Discharge The amount (volume) of water passing a point over a given length of time (or rate of flow). This is normally measured in litres per second or metres per second (cumecs).

Lag time The difference in time between the peak of rainfall and the peak of the flood.

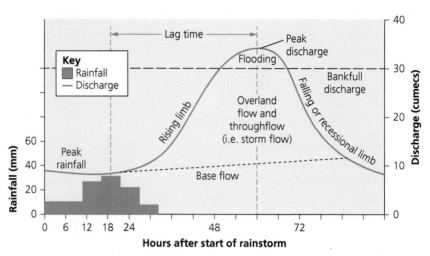

▲ Figure 2.18 A simple hydrograph

Factor	Influence on water cycle processes and storm hydrograph form
Climate	
Precipitation type and intensity	• Highly intensive rainfall is likely to produce overland flow, a steep rising limb and high peak discharge. Low-intensity rainfall is likely to infiltrate into the soil and percolate slowly into the rock, thereby increasing the lag time and reducing the peak discharge. • Precipitation that falls as snow is stored on the ground until it melts. Sudden, rapid melting can lead to high rates of overland flow and high peak discharge.
Temperature and evapotranspiration	• Not only does temperature affect the type of precipitation, it also affects the evaporation rate (higher temperatures lead to more evaporation and so less water reaching rivers). On the other hand, warm air can hold more water so the potential for high peak discharges in hot areas is raised. • Increased vegetation cover intercepts more rainfall and may return a proportion of it through transpiration, thereby reducing the amount of water reaching stream channels. The greater the return through evapotranspiration, the less water is able to reach stream channels, and therefore the peak of the hydrograph is reduced.
Antecedent soil moisture	• If it has been raining previously and the ground is saturated or nearly saturated, rainfall will quickly produce overland flow and a high peak discharge following a short lag time.
Drainage basin size and shape	• Smaller drainage basins respond more quickly to rainfall conditions. For example, the Boscastle (UK) floods of 2004 drained an area of less than 15 km². This meant that the peak of the flood occurred soon after the peak of the storm.
Drainage basin characteristics	
	• In contrast, the Mississippi River is over 3700 km long – it takes much longer for the lower part of the river to respond to an event that might occur in the upper course of the river. • Circular basins respond more quickly than linear basins, where the response is more drawn out.
Drainage density	• Basins with a high drainage density, such as urban basins with a network of sewers and drains, respond very quickly. • Networks with a low drainage density have a long lag time.

Porosity and impermeability of rocks and soils	• Natural rock and ground surfaces such as chalk and gravel are permeable and allow water to percolate downwards. This reduces the peak discharge and increases the lag time. • Sandy soils allow water to infiltrate, whereas clay is much more impermeable and causes water to flow over the land. • In contrast, impermeable urban surfaces generate more overland flow; this causes greater peak discharges.
Slopes	• Steeper slopes generate more overland flow, shorter lag times and higher peak discharges.
Vegetation type	• Broad-leaved vegetation intercepts more rainfall, especially in summer, and so reduces the amount of overland flow and peak flow and increases lag time. • In winter, deciduous trees lose their leaves and so intercept less: this is an important temporal control on water cycle flows and thus hydrograph characteristics.

▲ **Table 2.6** Factors affecting storm hydrographs

Urban hydrographs

The effect of urban development on hydrographs is to increase peak flow and decrease lag time (Figure 2.18). As we have seen (pages 46–7), this is due to an increase in the proportion of impermeable ground in a drainage basin as well as an increase in the drainage density (due to sewerage). Storm hydrographs also vary with a number of other factors (Table 2.6), such as basin shape, drainage density and gradient.

The encroachment onto the floodplain and river channel of embankments, reclaimed land and riverside roads can have a number of unintended hydrological consequences, including reduced channel width leading to increased height of floods in the restricted channel, and restricted flood discharge when bridges in the river temporarily dam the water and cause ponding upstream. The combined effects of all these changes is that the flow regime, the flood hydrology, the sediment balance and the pollution load of streams are all radically altered.

River regimes

In contrast to a storm hydrograph showing short-term water cycle dynamics, a river regime is the *annual* variation in the behaviour of a river. The character or regime of any stream or river is influenced by several variable factors:

- The amount and nature of precipitation (especially any seasonality in supply).
- The local rocks, especially porosity and permeability (a large basin may have heterogeneous geology, giving rise to complex catchment flow dynamics).
- The shape or morphology of the drainage basin, its area and slope (the sheer size of basins such as that of the River Colorado give rise to complex regime patterns with snowmelts occurring along different tributaries during different weeks and months).

 KEY TERM

River regime The variation in the discharge of a river over the course of a year. It is mainly determined by climate.

- The amount and type of vegetation cover (including any seasonal leaf-falls).
- The amount and type of soil cover (including consideration of agricultural practices, e.g. times of year when cropping exposes the soil and reduces the river basin's interception storage capacity).

On an annual basis the most important factor determining stream regime is climate. Figure 2.19 shows generalised river regimes for Europe.

- Notice how the regime for the Shannon at Killaloe (Ireland) has a typical seasonal discharge regime, with a clear winter maximum (as might be expected for a temperate climate in western Europe where winter brings an increased frequency of rain-bearing weather fronts).
- In contrast, Arctic areas such as the Gloma in Norway and the Kemi in Finland have a discharge peak in spring associated with snowmelt.

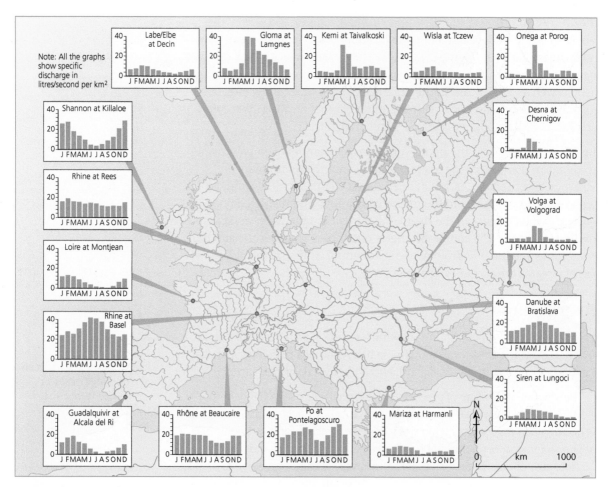

▲ **Figure 2.19** River regimes in different climatic zones of Europe

- Other regimes, such as the Po near Venice, have two main maxima linked with (i) the onset of autumn or winter rains (a feature of Mediterranean climates) and (ii) spring snowmelt from alpine tributaries.

As an extension task, you can use an atlas to identify further climatic and non-climatic factors (for example, relief) that might help explain why the river regimes shown in Figure 2.19 vary so much. You might try to analyse at least four regimes (one each from the west, east, north and south).

CONTEMPORARY CASE STUDY: RIVER REGIMES IN THE THAMES BASIN

The patterns of flow for the River Thames and its many tributaries are broadly similar; all show a maximum flow in winter and a lower flow in summer. However, on closer inspection there are important variations between the different hydrographs shown in Figure 2.20. This reflects marked differences in the characteristics of different streams and rivers on account of the varied character of the local places they flow through.

- For example, the River Ock, a small tributary of the Thames, mainly flows over impermeable clays. Clay is impermeable and, as a rule, increases overland runoff.

- By way of comparison, the River Kennet drains an area four times larger than the Ock, much of which consists of Tertiary rocks, sands and gravels. Although there is some abstraction of groundwater (and use of the water for agriculture and industry), the net impact of these artificial outputs from the water cycle is limited.

- In contrast, a third river, the Lambourn, flows mostly over chalk, and consequently has a much lower discharge (study the y-axes for the graphs: the scales differ and discharge of the Lambourn is typically less than one-quarter that of the Kennett). The river is maintained by groundwater, as rainwater soaks into the porous chalk rather than flowing over its surface. This results in very long lag times between periods of high rainfall and higher discharge.

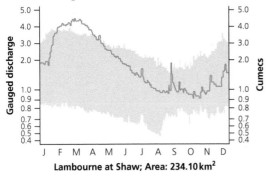

Lambourne at Shaw; Area: 234.10 km²

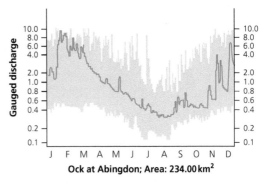

Ock at Abingdon; Area: 234.00 km²

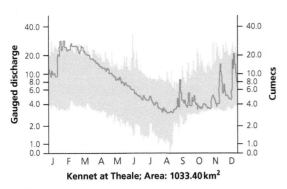

Kennet at Theale; Area: 1033.40 km²

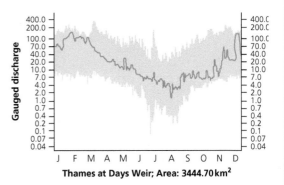

Thames at Days Weir; Area: 3444.70 km²

▲ **Figure 2.20** River regimes in the Thames Basin

Figure 2.21 compares the annual flow of water for the Coln (another tributary of the Thames) and the River Thames itself (in London) during 2016 and 2017.

- July 2017 was very wet, especially in the south of England; rainfall levels were at least 50 per cent higher than the long-term average value for this month of the year. However, the winter of 2016–17 and spring 2017 had been unusually dry.

- As a result, groundwater levels were very low, especially across much of the chalk and limestone areas of southern England, such as those that feed the Coln. The July flow for the Coln was the second lowest on record since 1963, just 50 per cent of the long-term average.

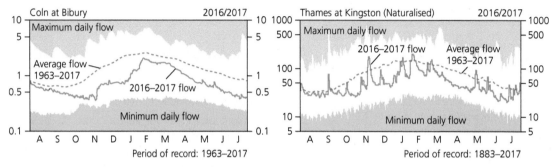

▲ **Figure 2.21** River flow for the Coln (tributary of the Thames) at Bibury and the Thames at Kingston, London

ANALYSIS AND INTERPRETATION

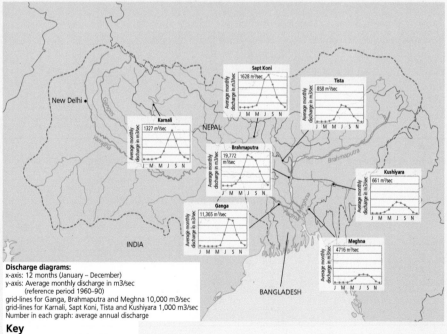

▲ **Figure 2.22** Average monthly discharge at selected stations in the Ganga-Brahmaputra-Meghna basin

Study Figure 2.22, which shows average monthly discharge at selected stations in the Ganga-Brahmaputra-Meghna basin.

(a) Analyse the temporal variations in maximum discharge for the rivers shown.

GUIDANCE

This question is asking you to analyse when the different rivers reach their maximum value. Is there a pattern? All seven stations show a maximum discharge in the summer. Most have their maximum in July but the Sapt Kosi at Baarakshetra and the Meghna at Bhairah Bazar have their peaks in August. The Ganga at Hardinge Bridge has a two-month peak (August–September).

(b) Suggest reasons for the variations in the temporal variations in maximum discharge.

GUIDANCE

Variations in the maximum discharge of the rivers is likely to relate to the onset of the monsoon rains, which are earlier in the east and in the south. For example, the Brahmaputra rises much earlier, and reaches its peak about one month before the Ganga. In addition, some of the smaller tributary rivers, such as the Karnali and the Sapt Kosi, begin rising in May. This is before the monsoon season, and so is likely to be the result of snowmelt. In contrast, the Ganga and the Brahmaputra are largely low altitude rivers so are not directly fed by snowmelt, although they are indirectly fed by snowmelt through their tributaries.

(c) Suggest why the annual discharges for these rivers vary so greatly in size.

GUIDANCE

You should use the map to provide some quantitative support for your arguments here. For example: the discharge of the Brahmaputra is enormous (over 19,000 cumecs), the Ganga (over 11,000 cumecs) and Meghna (over 4000 cumecs). Rivers combine to give the Ganga a discharge of over 34,000 cumecs in the Ganges Delta. These rivers drain an area of approximately 2000 km (east–west) and 1000 km (north–south). The drainage basin is much greater than for the Kushiyara (less than 700 cumecs) and the Tista (less than 900 cumecs).

⑤ Evaluating the issue

▶ *How far are human and physical factors responsible for changing water cycle flows in different drainage basins?*

Possible drainage basin contexts for studying changing water cycle flows

The focus of this chapter's final evaluation is *causality* in relation to changing water cycle flows, including flooding. Possible causes include:

- human factors, ranging from urbanisation and deforestation to the channelisation of water courses and even attempts to artificially increase rainfall inputs (such as cloud seeding)
- physical factors, including climatic, geological, ecological and edaphic (soil) influences. Sometimes, the operation of different physical factors will be interlinked: climate influences what vegetation can grow in an area, for instance, and is therefore viewed as a *dominant* or *key* factor influencing hydrological system operation.

This chapter has repeatedly demonstrated that water flows are subject to variation according to many different timescales, ranging from seconds (the sudden onset of a storm) to millennia (changes in cryosphere storage and melting linked to the onset or ending of ice ages, for example). Possible temporal contexts for exploring changing water cycle flows may therefore include the following:

- *Short-term changes.* Some desert rivers have very 'flashy' regimes, caused by irregular flash floods. So, too, do urban drainage basins where human activity is a dominant influence.
- *Seasonal changes.* Other rivers show more constant or seasonally defined flow variations,

e.g. those with monsoonal regimes, high latitude rivers (whose regimes are dominated by spring or summer snowmelt inputs) and Mediterranean flow patterns (winter flood, followed by summer drought). Urbanisation can amplify naturally occurring seasonal patterns such as these.

- *Complex annual flow patterns.* Some rivers have complex regimes and are influenced by many temporal patterns, e.g. the River Rhine's overall regime is influenced by (i) tributaries with a temperate regime (higher rainfall inputs in winter than summer), (ii) its alpine tributaries (whose peak flow is in spring or early summer), (iii) and tributaries whose flow has been heavily influenced by human activities (damming, abstraction, etc.).
- *Longer-term flow and regime changes.* Many rivers in southern Britain, such as the Evenlode and the Windrush, had past discharges that were much greater than those of today. During the periglacial phase in Britain, some 117,000–10,000 years ago, these areas had very cold winters and relatively short summers. Soil water froze, and so soils became impermeable with greatly reduced infiltration capacity. Seasonal snowmelt led to spring and early summer discharges perhaps 50 times greater than today. After the end of the periglacial period, conditions warmed, normal permeability resumed and seasonal water cycles became similar to today (higher in winter and lower in summer). These past changes occurred for natural reasons; in the future, anthropogenic climate change may bring long-term regime changes for many rivers.

Evaluating the view that *physical factors* are the most important cause of changing water cycle flows over time

Floods are perhaps the most obvious manifestation of changing water cycle flow patterns, and they are caused by many physical interactions, including floods, snowmelt, storms, high tides, storm surges, earthquakes and landslides creating 'quake lakes' (Figure 2.23).

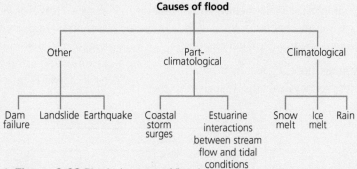

▲ **Figure 2.23** Physical causes of floods

Many floods occur as a result of prolonged heavy rain or due to a succession of storms. These can create saturated soils and promote overland flow. The very heavy snowfall in north-east USA during January 2018 led to widespread flooding. Intense seasonal rain, such as that in the south-east Asian monsoon, produces regular flooding, although the extent varies from year to year (Table 2.7).

Hurricanes regularly bring widespread flooding to many tropical and sub-tropical regions during each year's storm season (lasting from June to November in the northern hemisphere with a peak in September, and from November to April in the southern hemisphere with a peak around the end of February to early March). Hurricane Harvey brought over 1500 mm of rain to parts of southern USA in August 2017. Storm surges, which can accompany hurricanes, also cause coastal flooding. Hurricane Katrina had an accompanying storm surge of over 9 m, which over-topped the sea walls and levees in Louisiana and the southern USA in 2005. There is a view that hurricanes may be increasing in intensity on account of anthropogenic climate change: increasingly, human factors may therefore be playing a role in water cycle changes driven by hurricane activity.

In Nepal in 2015, a number of floods were caused by the over-topping of 'quake lakes' (lakes formed by the blockage of a valley by a landslide created due to earthquake activity). In such cases, a lake outburst eventually occurs through the debris that trapped water in the valley.

- This illustrates the interrelationship that can exist between different physical factors related to (i) tectonic systems and (ii) hydrological systems.
- The Indian plate and Eurasian plate push against each other until one gives way, releasing energy. The Moment magnitude (Mw) 7.8 earthquake and more than 450 aftershocks led to many deaths, much destruction and over 4000 landslides.
- By the end of the 2015 monsoon, many river valleys had been blocked by landslides, such as the Langtang valley. This led to the development of temporary lakes, which caused changes in the local hydrological cycle, including evaporation and channel storage.

Return period (years)	Approximate affected area (% of country)	Example of year in which this size flood was recorded
2	20	1999
5	30	2017
10	37	2004
20	43	1982
50	52	–
100	Around 60	1988
400–500	Around 70	1998

▲ **Table 2.7** Recurrence intervals and percentage of the country (Bangladesh) covered

- Some of these lakes would have been more longer lasting, but the landslides were broken up by local authorities in order to reduce the threat of flooding in the valley below. Thus, human activity was another factor to consider alongside the tectonic-hydrological interactions.

In many areas, flooding results from a combination of physical factors, such as in the Ganges valley where annual discharge peaks in Bangladesh are caused, in part, by monsoonal rainfall, snowmelt, tropical cyclones and storm surges. However, human causes are increasingly seen to be important in this context too, further suggesting that interrelations between different factors often bring the greatest changes to water cycle flows (Figure 2.24).

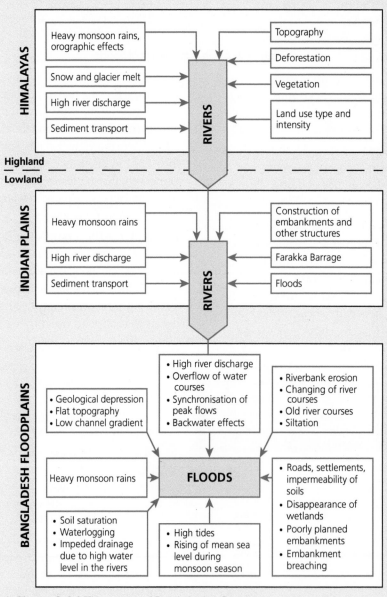

▲ **Figure 2.24** The causes of floods in the Ganges

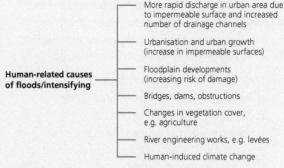

Human-related causes of floods/intensifying

- More rapid discharge in urban area due to impermeable surface and increased number of drainage channels
- Urbanisation and urban growth (increase in impermeable surfaces)
- Floodplain developments (increasing risk of damage)
- Bridges, dams, obstructions
- Changes in vegetation cover, e.g. agriculture
- River engineering works, e.g. levées
- Human-induced climate change

▲ **Figure 2.26** Flood intensifying conditions

- For example, the floods in Yorkshire, UK, in December 2015 are viewed by many as mainly the result of human activities in the catchment.
- For many years beforehand, campaigners in Hebden Bridge had canvassed the UK government to stop the artificial drainage and periodic burning of moors in the catchment (these are land-use practices associated with grouse hunting). However, while there are good reasons to think this may be true, this is contested.
- The damage inflicted on the moors and their deep vegetation reduces their capacity to function as a water store. Scrub, regenerating woodland, forested gullies, ponds and other features that harbour wildlife and hold back water have also been cleared.
- As a result, the argument goes, it is poor land use – rather than heavy rainfall – which should be blamed for changing water cycle flows. In its natural state, this catchment would have functioned well as a cascading system (see page 4) with multiple stores where the rainwater could have been held and slowly released to local rivers.

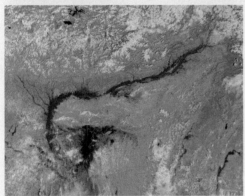

▲ **Figure 2.25** On 29 August 2014, the Moderate Resolution Imaging Spectroradiometer (MODIS) on NASA's Terra satellite captured the bottom image, which shows flooding along the Brahmaputra River and Tanquar haor, a large wetland region. The top image shows the same area on 8 September 2012, a more typical year. Both are false-colour images made from a combination of infrared and visible light. Water varies in colour from blue to black; vegetation is bright green; and bare ground is brown. This band combination makes it easier to spot changes in river dimensions.

Evaluating the evidence that *human factors* are the most important cause of changing water cycle flows over time

Turning now to human factors, a range of different direct and indirect actions and influences (Figure 2.26) may help alter drainage basin water cycle movements, notably urbanisation (see pages 46–9).

Increasingly, flood managers around the world recognise that flood prevention is needed as much as flood defence: land-use changes which can return local water cycle flows to their natural state, more than anything else, are seen as the key to managing flooding.

In another example of human intervention in water cycle flows, a whole series of new rivers have appeared in the central province of San Luis, Argentina.

- Until the 1990s, the Morro basin was a network of forests and grasslands, but has since been replaced by soy beans and maize (driven by growing economic demand within global systems for this vital staple crop – demand in China is particularly high).
- Soya now covers 60 per cent of Argentina's arable land. Unlike the deeply rooted forest originally present, which stored water and promoted evapotranspiration all year round, the soy bean has short roots and grows for only a few months each year.
- The result of this land-use change has been a rise in the water table, as well as an increase in overland flow and flooding. To try to reduce the impact of flooding, the provincial government passed an emergency law requiring land-owners to plant trees or water-consuming winter crops when the land is not being used for soy beans.

Turning finally to human factors operating on a global scale, there are signs that the global water cycle has already been affected by recent increases in the atmospheric carbon store attributable to human activity (see Chapter 5). Although it is hard to definitively prove that individual changing water cycle flows are caused by anthropogenic climate change, there are indications that global and local water systems are indeed being modified:

- In a warmer world, more evaporation takes place over the oceans – and what goes up must ultimately come down. Climate scientists believe rainfall patterns are changing in many parts of the world as the world's oceans warm (see pages 14–15).
- Climate change predictions for the UK suggest that the total amount of precipitation may not change but the pattern will become more seasonal. There could be an increased frequency of frontal rainfall in winter, and a greater probability of summer drought on account of reduced frontal rainfall.
- Extreme high-intensity rainfall in 2007 led to pluvial (surface water) flooding in some British cities, which caused £3 billion damage; high-intensity rainfall collected in low relief areas where homes were located (often in places far from a river or coastline). Higher temperatures in the future could mean an increased probability of high-intensity rainfall events.
- A consensus has emerged among UK meteorologists that the UK has entered a flood-rich period marked by more intense winter rainfall events. This is in line with IPCC climate change projections. Since the 1980s, river flooding has increased in size and duration during winter.
- In contrast, lower river discharges may become more common in southern England as a result of less frontal rainfall in summer.
- Although high-intensity convectional rain can be expected sometimes in summer, it is important to remember that it can lead to infiltration-excess overland flow. This means that depleted soil water and groundwater stores are not recharged. Under these conditions, river discharge will quickly return to a low level after a brief flashy response to the rainfall.

Arriving at an evidenced conclusion

What relative contribution do physical and human factors ultimately make to water flow variability on different timescales? As we have seen, the annual monsoon – driven by natural atmospheric processes – is perhaps the most significant aspect of water flow variability in Bangladesh. Moreover, the monsoon's interrelationship with other natural factors – such as gradient, vegetation cover, snowmelt, earthquakes and landslides – all contribute to the marked changes in water cycle flows experienced seasonally in some locations. But increasingly human factors are also affecting how much monsoon precipitation runs off the land as overland flow (Figure 2.25). Human activities in Bangladesh, as elsewhere, are varied and include land-use changes (deforestation) and urbanisation. There is a long history of human impacts in this region, although many of the impacts have intensified in the twentieth and twenty-first centuries as population has increased and standards of living have improved.

Human activities are placing more people at risk from flooding in every continent, not just Asia. Increased human habitation of floodplains increases vulnerability to flooding; so too do channel modifications and the construction of drains and ditches which often accelerate and

amplify water cycle movements in populated areas. But arguably the greatest current influence on changing water cycle movements – both globally and locally – is anthropogenic climate change. Most evidence suggests there are now clear links between changing carbon and water cycle movements. Taken together, a range of evidence suggests that Earth's climate is currently warming and changing; the US National Oceanic and Atmospheric Administration (NOAA) says the signs are 'unmistakable'. In 2015, global mean surface temperature (GMST) reached a new record high of +0.87°C relative to the 1951–80 average GMST, and the ten warmest years since 1880 have all been since 1998. These themes are returned to in Chapters 4 and 5.

 KEY TERM

Moment magnitude (Mw) The size of an earthquake measured by the amount of energy released.

Chapter summary

- At a global scale, the hydrological cycle is generally considered to be a closed system, whereas at a local scale, the drainage basin hydrological cycle is considered to be an open system. Both are driven by inputs of solar radiation but at the global scale there are no inputs (or outputs) of matter.

- There are many cascading stores and flows in the hydrological cycle; water cycle movements are complex and vary greatly in time and space. Their operation is dependent on a range of local environmental factors and the operation of, and interaction between, different physical systems.

- Urbanisation is an important land-use change which has major impacts on water cycle flows and stores, including infiltration and overland flow

rates. Recurrence intervals for floods of a certain magnitude decrease in urbanised areas.

- River regimes show the variations in a river's discharge over the course of a year, while storm hydrographs generally show the variation in discharge following a storm or short-term events.

- Reasons for changing water cycle flows in different contexts and at varying scales – ranging from Bangladesh to Hebden Bridge – are often complex, and there are opposing views on the overall significance of human activities.

- Ultimately, anthropogenic climate change is likely to alter water cycle flows on a planetary, continental and more local scales: evidence suggests changes are already occurring.

Refresher questions

1. What is meant by the following terms: water balance; river regime; Anthropocene?

2. Outline the evidence for sources of water existing deep inside the Earth's interior.

3. Using examples, outline the ways in which human activities have sometimes affected groundwater stores.

4. Outline the contribution different environmental factors make to the likelihood that overland flow will occur in a drainage basin.

5. Using examples, explain why (i) the infiltration capacity and field capacity vary for different soil types, and (ii) interception rates vary for different ecosystems.

6. Explain (i) the relationship between flood magnitude and flood frequency, and (ii) how flood frequency and flood magnitude vary with the degree of urbanisation in a catchment.

7. Describe how and explain why neighbouring urban and rural storm hydrographs may differ following a storm event.

8. Using examples, describe how and explain why river regimes vary from place to place within Europe.

Discussion questions

1. Working in pairs, discuss (i) the relative importance of reasons why flood risk has increased in large urban areas, and (ii) what could be done to reduce this risk.

2. In small groups, discuss the view that there is no such thing as a 'natural' drainage basin (where water cycle flows are unaffected by human activity).

3. In pairs, discuss the view that the key to successful flood management is detailed knowledge of local vegetation and soils.

4. In small groups, discuss the view that human activities have a greater impact on water cycle flows than natural processes do at the local level but not at the global level.

5. In pairs, design mind maps to show the multiple physical and human factors that influence the operation of (i) infiltration and (ii) overland flow. Use examples/statistics to add detail to your mind map.

FIELDWORK FOCUS

Water cycle dynamics lend themselves well to A-level Geography fieldwork and independent investigation titles. There is also plenty of scope to produce an investigation which broadly mirrors one of the more predictable management-themed essay titles associated with this topic, such as: an evaluation of a water management strategy's success; an assessment of why the strategy was needed; or an analysis of which players were involved.

A An investigation into the impact of new (housing) developments on drainage basin hydrology. This topic lends itself to primary data collection, such as observations of impermeable/permeable surfaces, vegetated/non-vegetated surfaces, measurements of hydrological processes and interviews with planners, conservationists and the general public. Environmental quality indices can be devised to examine the impact of the scheme. Secondary sources may include planning applications, environmental impacts statements and the local press. Stratified sampling may be used to ascertain the views of selected stakeholders such as long-term residents and those trying to get on the housing market/others hoping to benefit from the development.

B *An investigation into the soil and vegetation characteristics in a small area.* This may yield very detailed data relating to infiltration and/or interception. The work can be done over a short period or an extended period, e.g. looking for seasonal contrasts in precipitation or vegetation cover. It is best to focus on contrasting vegetation types (deciduous/coniferous trees, grass/shrub) and, if known, different soil types. Soils can be analysed in terms of their texture (i.e. the proportion of sand, silt and clay). Measurements can also vary in terms of gradient and proximity to a water body, and can also show different climatic conditions (low pressure/high pressure). Such a study would provide very useful data that would aid understanding of the water cycle, as well as providing for a detailed investigation.

C *How do storm hydrographs vary within a rural area or within an urban area?* Textbooks normally state that there are important variations between the storm hydrograph in an urban area and that of a rural area. However, there are important differences *within* rural areas and *within* urban areas. Reasons might include variations in geology, soil type, percentage of area that is impermeable, percentage of vegetation cover, the density of storm sewers and channels, for example. The presence of parks in urban areas is also important. Having collected reliable data, you would be able to use this in your examination papers, as well as producing a detailed investigation. One of the main considerations here is safety – you would need to work with a stream that is small enough to be safe, but large enough to show responses to the local conditions. Secondary data may be available from the Centre for Ecology and Hydrology and the Environment Agency.

Further reading

Alcantara-Ayala, I. and Goudie, A. (2014) *Geomorphological Hazards and Disaster Prevention*, Cambridge

Douglas, I. (1983) *The Urban Environment*, Arnold

Fei, H., *et al.* (2017) 'A nearly water-saturated mantle transition zone inferred from mineral viscosity', *Science Advances,* 7 June 2017, Vol. 3 (6), e1603024, pages 1–8

Hofer, T. and Messerli, B. (2006) *Floods in Bangladesh*, United Nations University Press

Smith, K., and Ward, R. (1998) *Floods: Physical Processes and Human Impacts*, Wiley

CHAPTER 3

Water security and sustainable water management

Water security is essential for human wellbeing: 'Thousands have lived without love, not one without water' (W.H. Auden). However, water security varies spatially as well as temporally, and human activities are making water scarcity more intense than ever before. This chapter:

- investigates the water budget and the causes of water deficits
- explores the reasons for water scarcity and water insecurity
- analyses sustainable water management strategies
- evaluates the extent to which water security can ever be fully guaranteed.

KEY CONCEPTS

Water budget The relationship between the inputs and outputs of a drainage basin.

Water security Having access to sufficient amounts of safe drinking water.

Water stress When per capita water supply is less than 1700 cubic metres per year.

 # The water budget and water deficits

▶ *What is meant by the water budget and how much does it vary spatially and temporally?*

The water budget (or balance) shows the relationship between the inputs and outputs of a drainage basin (see also page 26). It is normally expressed as:

precipitation = Q (discharge) = ET (evapotranspiration) +/– changes in storage (such as on the surface, in the soil and in the groundwater).

Figure 3.1 shows the water balance for a location in southern England, and how this affects availability of water in soils.

- Precipitation exceeds evapotranspiration between October and April, whereas evapotranspiration exceeds precipitation for the rest of the year.

- There are sufficient stores of water in the soil to be evaporated between May and June, whereas between July and September there is a water deficit. During this period, it may be important for farmers to irrigate their crops.

Figure 3.1 also illustrates how geographers recognise four distinct system states when analysing seasonal variations in soil water storage:

- **Soil moisture deficit** is the degree to which soil moisture falls below field capacity. In temperate areas, during late winter and early spring, soil moisture deficit is usually very low, due to high levels of antecedent precipitation and limited evapotranspiration in prior months.
- **Soil moisture recharge** occurs in autumn when precipitation exceeds potential evapotranspiration – there is some refilling of water in the dried-up pores of the soil.
- **Soil moisture surplus** is the period (typically the first few months of the calendar year in temperate regions) when soil is saturated and water cannot enter, and so flows over the surface.
- **Soil moisture utilisation** is the process operating in summer by which water is drawn to the surface through capillary action.

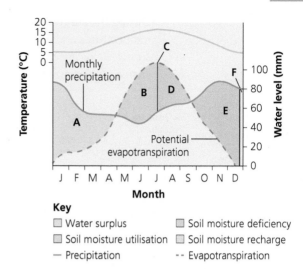

Key

☐ Water surplus	☐ Soil moisture deficiency
☐ Soil moisture utilisation	☐ Soil moisture recharge
— Precipitation	-- Evapotranspiration

A Precipitation > potential evapotranspiration. Soil water store is full and there is a soil moisture surplus for plant use. Runoff and groundwater recharge.

B Potential evapotranspiration > precipitation. Water store is being used up by plants or lost by evaporation (soil moisture utilisation).

C Soil moisture is used up gradually during this period. Any new precipitation is likely to be absorbed by the soil rather than produce runoff. River levels fall or rivers dry up completely.

D There is a deficiency of soil water (soil moisture deficit) as the store is used up and potential evapotranspiration > precipitation. Plants must adapt to survive, crops must be irrigated.

E Precipitation > potential evapotranspiration. Soil water store starts to fill again (soil moisture recharge).

F Soil water store is full, field capacity has been reached. Additional rainfall will percolate down to the water table and groundwater stores will be recharged.

▲ **Figure 3.1** Soil moisture status

Figure 3.2 provides another view of seasonal variability in catchment water storage. It shows the annual cycle of soil moisture in an agricultural region in Coshocton, Ohio. The cycle is generally representative of conditions in a humid, mid-latitude climate, in which there is a strong temperature contrast between summer and winter. In spring (March), the evaporation rate is low due to the low rates of energy input (reflected in low temperatures). Snowmelt and rainfall restores the soil moisture store to a surplus system state: for two months, the amount of water percolating through the soil and entering the groundwater keeps the soil pores fully occupied with water.

By May, increasing solar radiation leads to increased evaporation and vegetation growth leads to increased transpiration. This reduces the soil moisture to below field capacity, although it may be restored temporarily by occasional rainstorms. By the middle of summer, a moisture deficiency exists.

During the autumn, soil moisture begins to increase, due to reduced temperatures and less vegetation cover. At some stage during the winter/early spring, field capacity is again reached. This pattern of change is repeated each year: a good example of steady state equilibrium within a system.

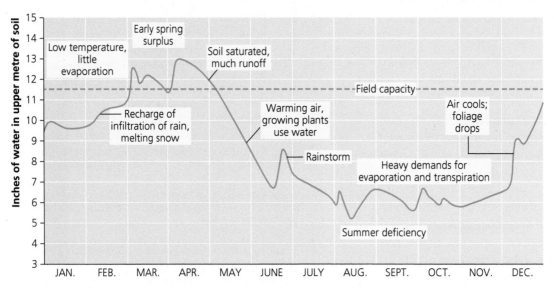

▲ **Figure 3.2** Annual cycle of soil moisture at Coshocton, Ohio, USA

In addition to the seasonal changes shown here, there are longer-term changes to consider for places where a state of dynamic (changing) equilibrium may exist (see page 10). For example, the water balance in southern England would have been very different during the last glacial period, and we would expect it to change in the next century as global warming leads to an increase in the UK's mean surface temperature and changing rainfall patterns.

Geologists and botanists have analysed old pollen samples extracted from soil and gravel deposits to show long-term climate changes. The results show conditions in the British Isles were much colder and wetter around 11,000–10,000 years ago (thereby favouring the growth of plant species whose pollen can still be found in soil deposits laid down at this time). Conditions became much warmer between 10,000 and 5000 years ago, with a warmer, drier climate. Evidence for this includes pollen extracted from the lake bed in Lough Neagh (Northern Ireland), along with hazel, oak, alder and elm pollen found in many bogs. In contrast, colder and wetter conditions are suggested by the birch and pine pollen found in earlier deposits. However, some of the changes in vegetation over time may also have been due to diseases, such as a fungus that attacked elm (similar to Dutch elm disease), and early human activities, notably farming, which replaced some species of trees with grasses. It is not always the case that pollen analysis invites a straightforward interpretation.

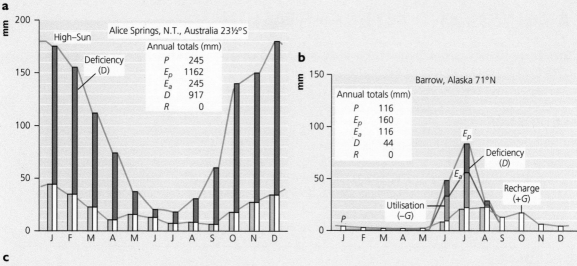

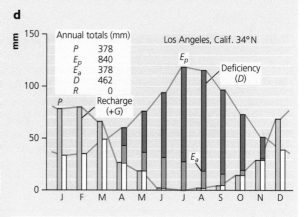

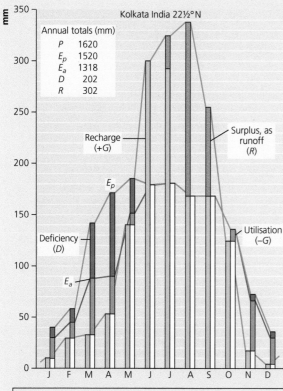

Key

E_a = Actual evaporation $+G$ = Recharge
E_p = Potential evapotranspiration $-G$ = Soil moisture utilisation
P = Precipitation D Soil moisture deficit/deficiency

▲ **Figure 3.3** Soil moisture budgets for contrasting locations: **a** Alice Springs, Australia, **b** Barrow, Alaska, **c** Kolkata, India, **d** Los Angeles, USA

▲ **Figure 3.4** World map showing the location of Alice Springs, Barrow, Kolkata and Los Angeles

ANALYSIS AND INTERPRETATION

Study Figure 3.3, which shows soil moisture budgets for four contrasting places which vary greatly in their climatic characteristics. Figure 3.4 shows the site locations.

(a) Distinguish between soil moisture deficiency and soil moisture utilisation.

GUIDANCE

This is a straightforward question essentially asking for two definitions, e.g. soil moisture deficiency occurs when the store of available water in the soil has been used up and when potential evapotranspiration is greater than precipitation. In contrast, soil moisture utilisation occurs when there is a store of water in the soil and it is available for use by plants or to be used in evaporation.

(b) Contrast the soil moisture budget for Alice Springs, Australia, and Barrow, Alaska.

GUIDANCE

This question asks us to contrast the differences in the soil moisture budget for two areas (but we are not asked to suggest reasons for the differences). It is sometimes easier to describe one station first, and then draw explicit contrasts when describing the second station.

For example:

- Alice Springs receives just 245 mm of rain annually, making it an arid location, but has a potential evapotranspiration of over 1100 mm. It has a year-round water deficiency totalling over 900 mm. The deficiency is greatest during December to January (summer in Australia), when it is approximately 125 mm/month. Whereas, during the winter, it is less than 10 mm/month.
- In contrast, Barrow has a more varied soil moisture budget. It is even more arid than Alice Springs. It receives just 116 mm of rain, making it a hyper-arid location. Monthly rainfall varies from a maximum of about 80 mm in July to less than 5 mm in the winter months. The period of soil moisture deficit is shorter than in Alice Springs, lasting about three months between June and August (summer). There is some soil moisture utilisation between July and August. Overall, there is a slight deficit of soil moisture (44 mm) and a slight recharge in October (in Alice Springs there was none).

(c) Suggest reasons for the differences in the soil moisture budget of Kolkata and that of Los Angeles.

GUIDANCE

In this answer you need to suggest reasons why the graphs differ. You may not have any prior knowledge of these places; instead, you are expected to make use of the resources provided and to apply your knowledge and understanding of geography to this novel situation. For example, possible points to make include the following:

- Kolkata has a very seasonal pattern to its soil moisture budget; so too does Los Angeles.
- Kolkata appears to have a monsoonal climate (the graph shows very high runoff and most likely flooding in summer, but dry in winter). There is soil moisture recharge each summer and sufficient soil moisture remains to allow for utilisation by plants (along with evaporation losses) throughout the autumn.
- Whereas Los Angeles is dry in summer but experiences rainfall in winter. In summer, a soil moisture deficit results from the lack of rainfall and high potential evaporation rates; but there is some recharge in winter due to the higher rains and lower evaporation rates.

Pearson Edexcel

AQA

OCR

② Drought, aridity, water scarcity and water insecurity

▶ *How do the causes of water scarcity and water insecurity vary?*

According to the seventh World Water Forum (2015), water stress and water scarcity are global challenges with far-reaching economic and social implications.

- Driven by increasing population, growing urbanisation, changing lifestyles and economic development, the total demand for water is rising. Urban centres, agriculture and industry all make increasing demands.
- World Water Forum experts hope in future to see greater evidence of nexus thinking in relation to water management. This means improved governance of water resources in ways which challenge traditional distinctions between competing economic sectors (such as agriculture or tourism) in order to help tackle water scarcity and insecurity issues. Co-operation is necessary not only between sectors or industries, but also between players at different hierarchical levels and scales (for example, from those working on the ground to the decision-makers, along with those working in different sectors, such as water, food and energy and different research fields).

Physical water scarcity and economic water scarcity

The level of water scarcity in a country depends on precipitation and water availability, population growth and affluence, demand for water, affordability of supplies and infrastructure.

 KEY TERMS

Water scarcity A lack of water due to either physical or economic reasons.

Nexus thinking Consideration of the complex and dynamic interrelationships between water, energy and food resource systems. Understanding of these interrelationships is essential if water and other natural resources are to be used and managed sustainably.

▲ **Figure 3.5** The impact of drought on soils

Where water supplies are inadequate, two types of water scarcity exist:

1 **Physical water scarcity**: where water consumption exceeds 60 per cent of the useable supply. To help meet water needs some countries, such as Saudi Arabia and Kuwait, have to import much of their food and invest in desalinisation plants.

2 **Economic water scarcity**: where a country physically has sufficient water to meet its needs, but requires additional storage and transport facilities. This means having to embark on large and expensive water-development projects, as in many sub-Saharan countries (Figure 3.5).

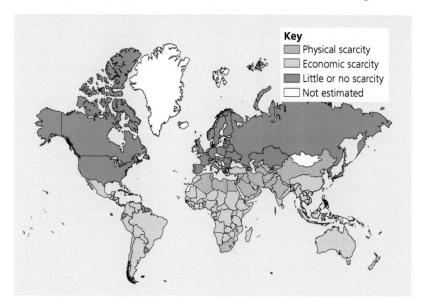

Key
- Physical scarcity
- Economic scarcity
- Little or no scarcity
- Not estimated

▲ **Figure 3.6** Physical and economic water scarcity

Drought and aridity

A drought is an extended period of unusually dry weather. The precise definition of drought varies from place to place, however.

- In the UK, for example, drought is defined as a 50 per cent deficit over three months, or a 15 per cent shortfall over two years (measured in relation to normal expectations for precipitation).
- In contrast, UNEP defines a drought as two or more years with rainfall substantially below the mean.

For example, the seasonal rains that usually fall between June and September in north-eastern, central and southern Ethiopia did not arrive in 2015. According to the UN, this was Ethiopia's worst drought in 30 years. Around 90 per cent of cereal production is harvested in autumn, after the summer-long rainy season, and the rest at the end of spring after the end of the short rainy season. The onset of this drought was linked to the El Niño weather system (see page 16), and resulted in a 90 per cent reduction in crop yields.

▲ **Figure 3.7** Drought in Ethiopia began in 2015, leading many people to rely on emergency food assistance

A large proportion of the world's surface naturally experiences dry or arid conditions. This is not the same as drought because low precipitation levels are a natural feature of the climate.

- Semi-arid areas are commonly defined as having rainfall of less than 500 millimetres per annum, while arid areas have less than 250 millimetres and in extremely arid areas the figure falls below 125 millimetres.
- In addition to low rainfall, dry areas typically have *variable* rainfall. As rainfall total decreases, variability usually increases. For example, areas with an annual rainfall of 500 millimetres have annual variability of about 33 per cent. This means that in such areas rainfall will typically range from 330 to 670 millimetres each year. This variability has important implications for vegetation cover, farming and the risk of flooding.

Water quantity and water quality issues

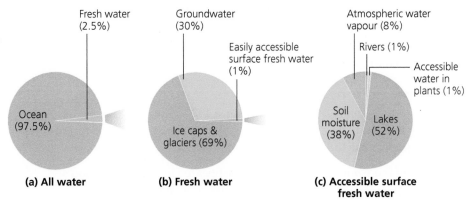

▲ **Figure 3.8** Availability of freshwater supplies

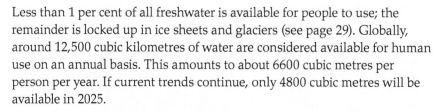

Less than 1 per cent of all freshwater is available for people to use; the remainder is locked up in ice sheets and glaciers (see page 29). Globally, around 12,500 cubic kilometres of water are considered available for human use on an annual basis. This amounts to about 6600 cubic metres per person per year. If current trends continue, only 4800 cubic metres will be available in 2025.

The world's available freshwater supply is not distributed evenly around the globe. Variations in annual supply may also be subject to marked seasonal differences in availability.

(see page 29)

- About three-quarters of annual rainfall occurs in areas containing less than a third of the world's population.
- Two-thirds of the world's population live in the areas receiving only a quarter of the world's annual rainfall. This is one reason which helps explain the enormous variability in per capita water footprints shown in Table 3.1. What other factors might help explain the range of values shown?

 KEY TERM

Water footprint A measure of the use of water by individual humans, nations and the amount needed to grow or manufacture products such as meat, textiles or steel.

The global challenge of water stress

When per capita water supply is less than 1700 cubic metres per year, an area suffers from water stress and is subject to frequent water shortages. In many such areas, per capita water supply is actually less than 1000 cubic metres per capita, which can create challenges for food production and economic development. In 2016, a total of 2.3 billion people lived in water-stressed areas. If current demographic and economic trends continue, water stress will affect 3.5 billion people – 48 per cent of the world's projected population – by 2025.

In many developing countries and emerging economies, access to adequate water supplies is increasingly threatened by the permanent loss or exhaustion of traditional sources, such as wells and seasonal rivers, for multiple reasons including conflict, dam building and climate change. Access may be worsened too by inefficient irrigation practices and lack of resources to invest in improvements. At present, 34 countries in Africa, Asia and the Middle East are classified as water-stressed. All but two of them, Syria and South Africa, are, as a result, net importers of grain. These countries collectively buy about 50 million tonnes of grain each year, about one-quarter of the total global trade volume.

Global patterns and trends in water use

The world's population has tripled since 1922. Humanity's water use has increased six-fold, however (Figure 3.9). Some rivers that once reached the sea, such as the Colorado in the USA, no longer do so. Moreover:

- half the world's wetlands have disappeared in the same period and 20 per cent of freshwater species are endangered or extinct

- water tables in many parts of the world are falling at an alarming rate and many important aquifers are being depleted.

Withdrawals of water for irrigation are nearly 70 per cent of the total demands placed on the water cycle by human activity, amounting to 2500 of 3800 cubic kilometres. Withdrawals for industry represent about 20 per cent; those for municipal use (drinking water and sewerage) account for the remaining 10 per cent.

Globally, human activity only makes use of around 10 per cent of renewable water resources. This is because of the highly uneven pattern of population distribution, however. Vast stores of water in high latitudes and continental interiors may go largely untouched; whereas concentration of population in warmer and drier regions – including the Mediterranean, North Africa, the Middle East and southern states of the USA – place very high demands on local water stores.

Currently, about 1.1 billion people lack access to safe water, 2.6 billion are without adequate sanitation and more than 4 billion do not have their wastewater treated to any degree. These numbers are likely to get worse in the coming decades. According to current predictions, by 2025, 4 billion people – half the world's population – will live under conditions of severe water stress. They will be disproportionately located in drier parts of Africa, the Middle East and South Asia. Disputes over scarce water resources could lead to an increase in armed conflicts.

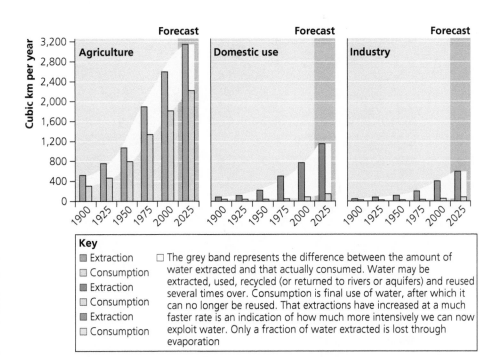

▲ **Figure 3.9** Trends in total global water use

Country	National water footprint (m³/year/per capita)
Brazil	2027
China	1071
Ethiopia	1167
India	1089
South Africa	1255
UAE	1000
UK	1258
USA	2842
Yemen	140
World average	1385

▲ **Table 3.1** National selected water footprints

Source: http://waterfootprint.org/en/water-footprint/national-water-footprint and IRIN

🔑 **KEY TERMS**

Green water The rainfall that is stored in the soil and evaporates from it; the main source of water for natural ecosystems, and for rain-fed agriculture, which produces 60 per cent of the world's food.

Blue water Renewable surface water runoff and groundwater recharge; the main source for human withdrawals and the traditional focus of water resource management.

Grey water Wastewater that has been produced in homes and offices. It may come from sinks, showers, baths, dishwashers, washing machines, etc., but does not contain faecal material.

Blackwater Wastewater that contains faecal material/sewage.

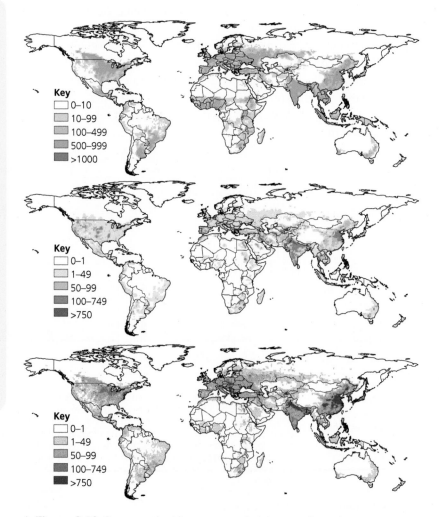

Key
0–10
10–99
100–499
500–999
>1000

Key
0–1
1–49
50–99
100–749
>750

Key
0–1
1–49
50–99
100–749
>750

▲ **Figure 3.10** Green-water, blue-water and grey-water footprints

ANALYSIS AND INTERPRETATION

Study Figure 3.10, which shows global variations in green-water, blue-water and grey-water footprints.

(a) Outline the differences between green water and blue water.

GUIDANCE

This question requires a brief answer only, as indicated by the command word 'outline'. Green water refers to water that is stored in the soil whereas blue water refers to renewable surface water runoff and groundwater recharge.

(b) Compare and contrast the global use of green water, blue water and grey water.

GUIDANCE

Part (b) asks for you to compare (look for similarities) and additionally make contrasts (look for differences). For example:

- There are very high uses of green, blue and grey water in eastern China, northern India and parts of south-east Asia such as the Philippines. However, what constitutes a 'high' rate on these maps varies between green water (>1000 mm), grey water (> 750 mm) and blue water (> 500 mm).
- In contrast, much of south central Canada and south Brazil-Argentina has a high green-water footprint but a medium grey-water footprint, and a negligible blue-water footprint.
- South-east Australia has a relatively high blue-water footprint but relatively lower grey-water and green-water footprints. Whereas contrastingly west Africa has a relatively high green-water footprint but a lower footprint for the other two types.

(c) Assess the value of the 'water footprint' concept used in Figure 3.9.

GUIDANCE

This answer requires a critical assessment of possible strengths and weaknesses of the concept of national water footprints. For example:

- A strength of the water footprint concept is the way it gives a clear indication of how much water the population of a particular country is consuming, helping us to identify possible inequalities and disparities in natural resource use occurring within global economic systems. The concept can also be applied at local, not just national levels, as these maps show. For example, water footprints are shown as being very high in parts of North America, Europe, India and China. However, not all local places make equal demands: western China and western USA have lower green-water footprints than their eastern parts (as you might expect given climatic conditions).
- A possible weakness in the water footprint concept is that many populations rely on so-called virtual (or embedded) water within food and goods that have been imported from outside the local area. This can consist of internal transfers (within a country such as the USA) or external transfers (international trade). It is important that water footprint analysis takes these flows into account (and it is unclear whether this is the case for the maps shown).
- Overall, the water footprint concept – as illustrated in these maps – is useful because of the way it draws attention to global, national and local disparities in natural resource use. However, it is important to know whether flows of embedded water between places are being taken into account as part of the analysis.

Patterns and trends in water availability and consumption

▶ *How does the availability and consumption of water vary spatially?*

Water supply depends on several factors in the water cycle, including the rates of rainfall, evaporation, the use of water by plants (transpiration) and river and groundwater flows.

- Water is unevenly distributed over the world, and over 780 million people do not have access to clean water (Table 3.2). The global population is likely to increase to 9 billion by 2050, which, combined with changes in diet, will increase demand for water, particularly in Africa where the population is predicted to nearly triple to 4 billion people by 2100. Moreover, the increased demand for water for hydro-electric power will further strain the Earth's water resources.
- In studies of water availability and consumption, an important distinction exists between 'improved' and 'unimproved' water sources (Table 3.3). Improved water sources are those where the source is protected from outside contamination, especially faecal material. Unimproved sources are those in which the quality may fail to reach safe standards. Bottled water may be classified as 'improved' or 'unimproved' depending on where the source of the water originated (and whether it was treated in any way).

Table 3.2 summarises the numbers and proportions of the people in developing countries lacking access to water and adequate sanitation in the years 1990 and 2015.

Access to safe water	1990	2015
World	76	91
Central Europe and the Baltics	92	99
East Asia and Pacific	71	94
Europe and Central Asia	95	98
European Union	98	100
Latin America and Caribbean	85	95
Middle East and North Africa	87	94
North America	99	99
South Asia	72	92
Sub-Saharan Africa	47	68

▲ **Table 3.2** Global water supply

Source: World Bank, https://data.worldbank.org/indicator/SH.H2O.SAFE.ZS

The following technologies are considered to be improved	The following technologies are considered unimproved
• Water supply	• Water supply
• Household connection	• Unprotected well
• Public standpipe	• Unprotected spring (Figure 3.11)
• Borehole	• Vendor-provided water
• Protecting dug well	• Bottled water
• Protecting spring	• Tanker truck provision of water
• Rainwater collection (Figure 3.10)	
• Bottled water	

▲ **Table 3.3** Water supply technologies considered to be improved and unimproved

Seasonal water availability and consumption

Throughout much of the developing world, freshwater supply comes in the form of seasonal monsoon rains (see page 58). These rains often run off too quickly for efficient use. India, for example, gets 90 per cent of its rainfall during the summer monsoon season – at other times rainfall over much of the country is very low. Because of the seasonal nature of the water supply, many developing countries can use no more than 20 per cent of their potentially available freshwater resources. Water supplies can also vary from year to year; for example, natural phenomena such as El Niño can lead to significant annual or decadal differences in rainfall in the southern Pacific Ocean, affecting south-east Asian and South and Central American countries alike. The study of water use issues – in common with other water cycle topics included in this book – therefore invites critical discussion of the way physical processes vary in the way they operate over many different timescales.

Urban and rural water availability

Urban areas are usually better served than rural areas (although this can change; for instance, during periods of conflict when urban supplies are cut off). Many piped water systems, however, do not meet water quality criteria, leading more people to rely on bottled water bought in markets for personal use (as in major cities in Colombia, India, Mexico, Thailand, Venezuela and Yemen). Consumption of bottled water in Mexico is estimated at more than 15 billion litres a year.

▲ **Figure 3.11** Water collection in Antigua

▲ **Figure 3.12** An unprotected lake, Chalumna, Eastern Cape, South Africa

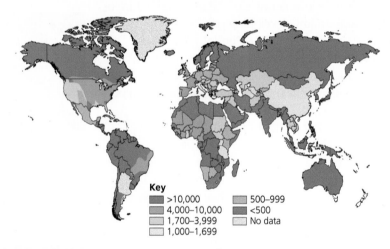

Key
>10,000
4,000–10,000
1,700–3,999
1,000–1,699
500–999
<500
No data

▲ **Figure 3.13** Annual renewable water (m³/person/year)

New water availability and consumption challenges

In future years, water availability is likely to decrease in many regions at the same time as demand is rising. For example, 300 million people in sub-Saharan Africa already live in a water-scarce environment (Figure 3.13) and an increase of 2°C could see crop yields of maize, sorghum and wheat decline by 17–22 per cent, and a reduction of between 40 and 80 per cent of current crop-growing areas. Increased temperatures could lead to increased evapotranspiration losses, and therefore reduced water availability, potentially causing a decline in crop yields. Central and southern Europe are predicted to get drier due to climate change. In a high-emissions climate change scenario (see page 15), mean average surface temperature may rise by

7–8° in parts of Spain and Portugal, radically affecting water cycling in regions such as the Ebro Delta, as evaporation rates increase markedly.

Developed countries are tending to maintain or increase their consumption of water resources, although an increasing proportion of water is embedded in agricultural and manufactured products imported from elsewhere due to the growth of complex global trading systems in recent decades. The average North American or Western European adult consumes $3\,m^3/day$, compared with around $1.4\,m^3/day$ in Asia and $1.1\,m^3/day$ in Africa.

In some cases, it is not just the international, or transboundary, transfer of water through traded food and products that reduces local water availability in some places. The loss of local people's water stores because of land grabs also plays a role. For example, several years ago Saudi Arabia cut its own domestic production of cereals by 12 per cent but compensated for this via a series of land grabs in Ethiopia, Sudan, South Sudan, Mali, Mauritius, Morocco and Tanzania (including a 2014 lease of 10,000 acres of land in the Gambella region of Ethiopia, to help secure Saudi Arabian food supplies). However, this led to food insecurity, landlessness and reduced access to water for local Ethiopians. Ultimately, some conflict also occurred – 12 Saudi Arabian workers were murdered by rioting Ethiopians.

Finally, various developmental trends are increasing the pressure to manage water more efficiently in order to meet mounting consumer demands. These include:

- world population growth – some estimates say it may eventually peak at 11 billion
- the growing middle class – increasing affluence leads to greater water consumption, for example as more people take showers and baths, or own appliances like dishwashers
- the growing demand for tourism and recreation, for example golf courses, water parks, swimming pools (Asian tourist markets are growing rapidly in size, putting greater stress on water resources)
- urbanisation – urban areas require significant investment in water and sanitation facilities to get water to people and to remove waste products hygienically.

Virtual water consumption

The concept of virtual (or embedded) water refers to the way in which water is transferred from one country to another through its exports (Figure 3.13). These exports may be foods, flowers or manufactured goods, for example.

- A can of cola contains 0.35 litres of water, yet it requires an average of 200 litres to grow and process the sugar contained in that can.
- It takes 2900 litres to 'grow' a cotton shirt and 8000 litres to produce a pair of leather shoes; that is, the amount of water required to grow, feed, support and process that fraction of a cow that makes just one pair of leather shoes.

> **KEY TERM**
>
> **Land grab** The purchase or leasing (rental) of large areas of land, mostly for the production of food for export, by TNCs, foreign governments and domestic companies/government. It mainly occurs in countries with weak national governance.

Trade in goods with a high water footprint allows countries with limited water resources to 'outsource' their water from countries with greater water resources. It also allows a country to reduce the use of its own water resources by importing 'water-hungry' goods. For example, Mexico imports maize, and thereby saves 12 billion cubic metres of water for its own domestic use each year. Figure 3.13 shows the global network of interconnected places that supply the UK with goods and food which have a high embedded water value. As you can see, the water demands of just one country turn out to be vast.

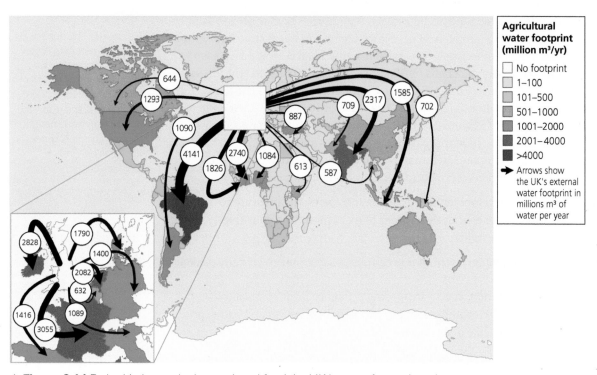

▲ **Figure 3.14** Embedded water in the goods and food the UK imports from other places

People in developed countries like the UK tend to have a more varied diet, with higher proportions of meat and dairy products; hence the large amounts of virtual water they consume. This is increasingly true of the majority of people in some emerging economies too, including Brazil and China.

- A much larger amount of water is required to produce a kilogram of meat or dairy product than it is to produce a kilogram of crop. This is partly because the crops are fed to the animals, which require additional water for movement, respiration and hydration.
- 'Luxury' drinks such as coffee, wine and fruit juice have a high embedded water value (it takes 1100 litres, 960 litres and 950 litres of water to grow the crops needed to make one litre of coffee, wine and apple juice respectively).

- Thus, purely in terms of diet, the water footprint of someone in a developed country is higher than that of someone in a developing country. In addition, a much higher proportion of the food consumed in developed countries comes from developing countries, and so there is a transfer of virtual water from the producer to the consumer.

Many trade flows within global systems are therefore, in essence, hydrological transfers between distant drainage basins.

CONTEMPORARY CASE STUDY: CONTRASTING WATER SUPPLY ISSUES IN THE USA AND SOUTH SUDAN

The USA is a huge user of water. However, the western states of the USA, covering 60 per cent of the land area with 40 per cent of the total population, receive just 25 per cent of the country's mean annual precipitation. Yet each day the west uses as much water as the east. This is the area of the USA most vulnerable to water shortages.

- The south-west has prospered due to a huge investment in water transfer schemes. Hundreds of aqueducts take water from areas of surplus to areas of shortage.

- Although much of the south-west is desert or semi-desert, large areas of dry land have been transformed into fertile farms and sprawling cities. It began with the Reclamation Act of 1902 which allowed the building of canals, dams and hydro-electric power systems in the western USA.

- California has benefited most from this investment in water supply. Seventy per cent of runoff originates in the northern one-third of the state but 80 per cent of the demand for water is in the southern two-thirds. Agriculture uses more than 80 per cent of the state's water, though it accounts for less than a tenth of the economy.

▲ **Figure 3.15** Hoover Dam

▲ **Figure 3.16** Swampland in South Sudan

The 2333 km long Colorado river is an important source of water in the south-west. Over 30 million people in the region depend on water from the river. Completed in 1936, the Hoover Dam and Lake Mead marked the beginning of the era of artificial control of the Colorado. Despite the interstate and international agreements (between the USA and Mexico), major problems over the river's resources have arisen because population has increased along with rising demand from agriculture and industry.

The $4 billion Central Arizona Project (CAP) is the latest, and probably the last, big money scheme to divert water from this great river. Since CAP was completed in 1992, 1.85 trillion litres of water a year has been distributed to farms, Indian reservations, industries and fast-growing towns and cities along its 570 km route between Lake Havasu and Tucson. However, providing more water for Arizona has meant that less is available for California.

The Colorado Delta is now dry for much of the year and has become starved of sediment. Fish and bird populations have declined considerably. Recycling methods have resulted in some farmers having highly saline water, which can damage their crops.

South Sudan, in contrast, is a large landlocked country in east-central Africa (it is about one-third the size of western Europe). According to WASHWatch, over 5 million people lack access to improved water supplies. Water supply in Sudan has numerous challenges. Naturally, rainfall varies, from 400 mm in the north to around 1600 mm in the south. The dry season lasts from around November to June. The country's water resources are unevenly distributed both spatially and temporally. Around 50 per cent of the flow of water into the White Nile is lost in the Sudd wetlands, the largest swamp in the world.

- Access to safe water has been reduced due to ongoing conflict in South Sudan, the legacy of some 40 years of civil war, mass internal displacement of people, and, in some cases, the deliberate destruction of water infrastructure.

- Current pressures include rapid urbanisation and the influx of migrants looking for employment. Both of these are putting a strain on water resources in urban areas. Some financing of water projects in urban areas have been made by US, German and Japanese firms.

- The proportion of water generated from rainfall is erratic and prone to long dry seasons and drought. In dry years, rainfall may be down 15 per cent – this variability, coupled with limited investment in storage, makes South Sudan vulnerable to droughts and floods. South Sudan is believed to have substantial aquifers, but some of these have been affected by dumping of waste, salinity and oil.

	South Sudan	USA
Population (m)	13	326
Population growth rate (%)	3.83	0.81
PPP (US$)	1500	58,200
% of population below the poverty line	50.6	15.1
% employed in agriculture	72	0.7
% employed in industry	7	20.3
% employed in services	21	79.0
% with improved drinking water Urban	66.7	99.4
Rural	56.9	98.2

▲ **Table 3.4** Comparing the USA and South Sudan

In 2015, the World Food Programme warned about the onset of drought in South Sudan, and other neighbouring countries, due to a developing El Niño event. As a result, the length of the rainy season grew shorter and so too did crop yields. Five million people in South Sudan were left with insufficient food caused by a combination of drought and war. The crisis lasted until June 2017, when the UN declared that the famine was over. Unfortunately, continued political instability, population growth and climatic change mean that water insecurity – and associated food shortages – may remain a recurring and potentially intensifying problem.

Water challenges and sustainable solutions

▶ *Why do some people and places lack a reliable and safe water supply, and how can water challenges be tackled?*

The most obvious uses of water for people are drinking, cooking, bathing, cleaning, and – for some – watering family food plots. In addition to important water quantity issues affecting these uses, water also needs to be of an adequate *quality* for consumption. However, in developing countries too many people lack access to safe and affordable water supplies and sanitation. The World Health Organisation (WHO) estimates that around 4 million deaths each year result from cholera, hepatitis, malaria and other water-related parasitic diseases. Water quality may be adversely affected by organic waste from sewage, fertilisers and pesticides from farming; also by heavy metals and acids from industrial and transport processes.

Sufficient clean water is clearly essential to everyone's wellbeing. Yet over one billion people do not have access to safe and affordable drinking water while 2.5 billion people – nearly two-fifths of the world's population – live in conditions lacking adequate sanitation. The vast majority of these people are in the poorest developing countries, along with urban slums and rural areas in some emerging economies. In these contexts, insufficient supplies of water and sanitation disproportionately affect women, children and the poor. Each year about 4 million people die of waterborne diseases, including 2 million children who die of diarrhoea (most deaths from water-related diarrhoea are among children under fifteen). Accordingly, the United Nations has identified water use as a priority for international aid and access to water is now recognised as a key issue in development.

More than 800 million people – 15 per cent of the world population – are malnourished, due in part to insufficient water for crops. This figure could increase further in the future: nearly 3 billion people could face severe shortages of fresh water by 2025 if the world keeps using water at today's rates. Urban populations are, in theory, better placed than rural populations to avoid health problems linked with water supply issues. Many urban piped water systems do not meet water quality safety benchmarks, however, leading more people to rely on bottled water. This is traded in vast quantities in markets of major cities in Colombia, India, Mexico, Thailand, Venezuela and Yemen. As a result, in some cases the world's poor actually pay more for their water than the rich. For example:

- In Port-au-Prince, Haiti, surveys have shown that households connected to the water system typically paid around US$1.00 per cubic metre, while unconnected customers forced to purchase bottled water from mobile vendors paid from US$5.50 to a staggering US$16.50 per cubic metre.
- Urban residents in the United States typically pay US$0.60 per cubic metre for municipal water of excellent quality. Whereas in the slums of Peru's megacity Lima, poor families pay vendors roughly US$3.00 per cubic metre for bottled water.

In contrast to the poor water quality experienced by some populations in the developing world, water supplies in developed countries are subject to a high level of regulation and far stricter governance in order to protect water quality. In the UK, the Environment Agency (EA) works in close partnership with water companies, farmers, businesses and environmental organisations to achieve this goal. The EA has legal powers to impose significant fines on businesses or individuals found guilty of serious offending (i.e. whose actions have a significant negative impact on water quality).

Competition for water at varying scales

As water becomes scarcer, competition for it increases. This happens not just between economic sectors within individual countries, but between nations too.

Rapidly growing cities and industries increasingly depend on irrigated agriculture as a source of water. However, a cubic metre of water used in China's urban industries generates more jobs and about 70 times more economic value than the same quantity used in farming. In this context, should agriculture or industry be viewed as the priority user within China?

At a larger scale, competition is also increasing between countries, as populations continue to grow in some of the world's water-short regions. The Mekong is south-east Asia's largest river. It is the world's twelfth longest and twenty-first in the size of its basin. The Mekong's hydrology remained almost untouched until 1993 when the first dam on the river, at Man Wan in China, was completed. Since then, population growth and economic growth throughout south-east Asia have put enormous strain on the Mekong River and its drainage basin (Table 3.5), particularly given its enormous hydro-electric potential.

	Population (m)	Population growth, 2018 (%)	Economic growth, 2018 (%)
Cambodia	16	1.52	6.9
China	1379	0.41	6.8
Laos	7	1.51	6.9
Myanmar (Burma)	55	0.91	7.2
Thailand	68	0.30	3.7
Vietnam	96	0.93	6.3

▲ **Table 3.5** Population, population growth and economic growth in the Mekong region, 2018

Plans for international sharing of the Mekong's water resources have been in development for decades.

- In 1970, the first Indicative Basin Plan of the Mekong was a first attempt to develop an international governance framework for what the plan called 'south-east Asia's river of promise'.
- In 1995, the Mekong Agreement on the Co-operation for the Sustainable Development of the Mekong River basin was signed. This was later described by the World Commission on Dams as 'a good model for developing countries to follow in international River Basin Management' on account of the way the agreement aimed to ensure benefits would be shared by different countries in the Mekong drainage basin, thereby creating opportunities for development while also promoting stability and peace in the region.

River basin/countries	Population 1999	Projected 2025 population	Change (%)
Aral Sea Kazakhstan, Kyrgyzstan, Tajikstan, Turkmenistan, Uzbekistan	56 million	74 million	+32
Ganges Bangladesh, India, Nepal	1137	1631	+43
Jordan Gaza, Israel, Jordan, Lebanon, Syria, West Bank	34	58	+71
Nile Burundi, DR Congo, Egypt, Eritrea, Ethiopia, Kenya, Rwanda, Sudan, Tanzania, Uganda	307	512	+67
Tigris-Euphrates Iraq, Syria, Turkey	104	156	+50

▲ **Table 3.6** Populations in areas of potential water conflict and projected population growth

Sustainable water management strategies

How can the water cycle of different drainage basins be best managed in ways that meet the growing needs of different user groups and places for adequate supplies of safe drinking water? Numerous sustainable water management strategies exist, including irrigation, desalination and storage in dams. Each of these is now evaluated in turn.

Traditional irrigation strategies

People have irrigated crops since ancient times: there is evidence of irrigation in Egypt dating back nearly 6000 years. Water for irrigation can be diverted from hydrological surface stores, such as lakes, dams, reservoirs and rivers, or from subsurface groundwater stores. Types of irrigation range from total flooding (as is used for paddy fields), to drip irrigation, where precise amounts are measured out and allocated to each individual plant (Figure 3.17). Irrigation may, however, bring unwanted consequences such as salinisation.

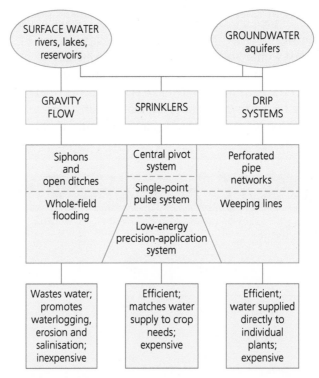

▲ **Figure 3.17** Types of irrigation

Irrigation occurs in countries at all stages of economic development (Table 3.7): only the methods differ. For example, large parts of the USA and Australia are irrigated. The advent of diesel and electric motors in the mid-twentieth century led for the first time to systems that could pump groundwater out of aquifers faster than it was recharged. This has led in some regions to a loss of aquifer capacity and decreased water quality. In Texas, USA, irrigation has reduced the water table by depths of up to 50 metres. This has occurred because groundwater has been extracted at a rate faster than it can replenish, thereby lowering the water table. In contrast, in the fertile Indus Plain in Pakistan, irrigation has actually raised the water table by as much as 6 metres since 1922 and caused widespread salinisation as a result. This occurs in hot areas where the addition of irrigation water raises a water table that is close to the surface. Soluble salts may be brought to the surface. Water evaporates and a saline crust may be left (salinisation).

Irrigation also contributes to a reduction in ground albedo (reflectivity) by as much as 10 per cent. This happens when a reflective sandy surface is replaced by dark green crops. Large-scale irrigation in semi-arid areas, including the High Plains of Texas, has been linked with increased rainfall, hailstorms and tornadoes as a result of changing water cycle fluxes and flows. Under natural conditions semi-arid areas have sparse vegetation and dry soils in summer. Once irrigated, however, these areas instead have moist soils in summer and complete vegetation coverage. Evapotranspiration rates increase markedly, resulting in greater amounts of summer rainfall (see page 65–6).

New approaches to irrigation

At the start of the twenty-first century, the need for new approaches to irrigated farming is clear. One of the starkest indicators of the problems associated with irrigation is the depletion of underground aquifers. Over the last 30 years, the number of groundwater wells worldwide has expanded at an exponential rate. Groundwater depletion has changed from an often small-scale and isolated phenomenon to an extensive and pervasive challenge affecting entire cropland regions.

There is a long list of measures that can help improve irrigation water productivity and reduce stress on groundwater stores. One of these is drip irrigation, which consists of a system of plastic tubes installed at or below the ground surface that deliver water to individual plants and minimises wasteful system outputs through either evaporation or overland flow (Figure 3.16). The water, which can also usefully be enhanced with fertiliser, is delivered directly to the roots of plants, so that there is very little loss via other water flow mechanisms.

Drip irrigation can achieve 95 per cent efficiency compared with 50–70 per cent for conventional flood irrigation systems where up to half of all water inputs are wasted. In surveys carried out across the USA, Spain, Jordan, Israel and India, drip irrigation has been shown to cut agricultural water use by between 30 per cent and 70 per cent while also increasing crop yields by 20–90 per cent (thereby almost doubling productivity in the most successful cases). Unfortunately, drip irrigation so far accounts for only about 1 per cent of all irrigated land worldwide. Much wider adoption is needed.

Irrigated area	Million km²
China	690
India	667
USA	264
Pakistan	202
Iran	95
Indonesia	67
Mexico	65
Thailand	64
Brazil	54
Bangladesh	53
Turkey	52
Vietnam	46
Russia	43
Uzbekistan	42
Italy	39
Spain	38
Egypt	36
Iraq	35
Afghanistan	32
Romania	31

▲ **Table 3.7** Irrigated land worldwide: the 20 countries with the largest amount of irrigated farmland

Source: *CIA World Factbook*, 2018

▲ **Figure 3.18** Drip irrigation, United Arab Emirates

Irrigation method	Typical efficiency (%)	Water application (mm) needed to add 100mm to root zone	Water savings over conventional furrow irrigation (%)
Conventional furrow	60	167	–
Furrow with surge valve	80	125	25
Low-pressure sprinkler	80	125	25
Low energy precision application (LEPA) sprinkler	90–95	105	37
Drip	90–95	105	37

▲ **Table 3.8** Efficiencies of selected irrigation methods, Texas High Plains

To date, the benefits of modern irrigation technology have not always been shared equally among the world's agricultural water users. The cheapest way to use groundwater in large amounts, for example, is to use a diesel pump on a tubewell. These generally cost about US$400, putting them out of the reach of small farmers in sub-Saharan Africa. There is, however, an array of irrigation technologies that are low cost, affordable and acceptable to small farmers in developing countries (Table 3.9). An example of one that has been very successful is the treadle pump, a human-powered irrigation system. It is very similar to bicycles found in gyms, in which someone pedals on a machine while grasping the handles. In this case, the pedalling causes groundwater to be sucked up into a cylinder.

Technology or method	General conditions where appropriate	Examples
Cultivating wetlands, delta lands, valley bottoms; flood-recession cropping; rising flood cropping	Seasonally waterlogged floodplains or wetlands	Niger and Senegal river valleys; fadama of Nigeria; dambos of Zambia and Zimbabwe; other parts of sub-Saharan Africa
Treadle pump, rower pump; pedal pump; rope pump; swing basket; Archimedean Screw; shaduf or beam and bucket; hand pump	Very small (less than 0.5 hectares) farm plots underlain by shallow groundwater or near perennial streams or canals in dry areas or areas with a distinct dry season	Eastern India; Bangladesh; parts of south-east Asia; valley bottoms, dambos, fadama, and other wetlands of sub-Saharan Africa
Persian wheel; bullocks and other animal-powered pumps; low-cost mechanical pumps	Similar to those above, but where the average size of farm plots is roughly 0.5–2.0 hectares	Those above, in addition to parts of North Africa and Near East
Various forms of low-cost micro-irrigation, including bucket kits; drip systems; pitcher irrigation; as well as micro-sprinklers	Areas with perennial but scarce water supply; hilly, sloping or terraced farmlands; tail-ends of canal systems; can apply to farms of various sizes, depending on the micro-irrigation technique	Much of north-west, central and southern India; Nepal; Central Asia, China, Near East; dry parts of sub-Saharan Africa; dry parts of Latin America
Tanks; check dams; percolation ponds; terracing; bunding; mulching; other water-harvesting techniques	Semi-arid and/or drought-prone areas with no perennial water source	Much of semi-arid South Asia, including parts of India, Pakistan and Sri Lanka; much of sub-Saharan Africa; parts of China

▲ **Table 3.9** Low-cost irrigation methods for small farmers

Desalination: the ultimate 'technological fix' for water challenges?

Desalination removes salt from seawater. More generally, desalination may also refer to the removal of salts and minerals from seawater and saltmarshes to produce fresh water fit for human consumption (potable water) including irrigation use.

Due to high energy input, the financial costs of desalinating seawater are generally high still. But alternative water sources are not always available and around 4 per cent of the world's population are already dependent on desalinated water to meet their daily needs. By 2025, it will most likely have risen steeply to around 14 per cent.

- In 2015, there were more than 18,000 desalination plants operating worldwide, producing almost 90 million cubic metres of water per day for 300 million people. The world's largest single desalination project is Ras Al-Khair in Saudi Arabia, which produced over one million cubic metres per day in 2014.
- In Israel, over 40 per cent of domestic water comes from seawater desalination, the largest contribution to any one country. As recently as 2004, Israel relied entirely on groundwater and rain; it now has four seawater desalination plants running, of which Sorek is the largest. By 2050, seawater desalination is expected to account for 70 per cent of Israel's water supply.
- Desalination is becoming more popular among wealthy countries that can afford it. Australia, Singapore and several countries in the Persian Gulf are already heavy users of seawater desalination, and California is also starting to embrace desalination.

The main criticism regarding desalination and the use of reverse-osmosis technology is that it costs too much. The process uses a great deal of energy to force salt water against membranes that have pores small enough to let fresh water through while holding salt ions back. However, new technology is continually bringing the cost of desalination down. Currently, the Sorek plant produces the cheapest water from seawater in the world at a cost of around US$0.60 per cubic metre.

The construction of large dams and their hydrological implications

The number of large dams (more than 15 metres high) being built around the world is increasing rapidly; there are almost two new completions every day! The advantages of dams are numerous. They include flood and drought control (Figure 3.17), irrigation, hydro-electric power, improved navigation, recreation and tourism. On the other hand, there are numerous costs. For example, these include water losses through evaporation, salinisation, displacement of population, drowning of archaeological sites, seismic stress, channel erosion (clear water erosion) below the dam, silting upstream of the dam and reduced fertility (sediment decline) downstream from the dam.

▲ **Figure 3.19** Water storage in the Paphos Dam, Cyprus

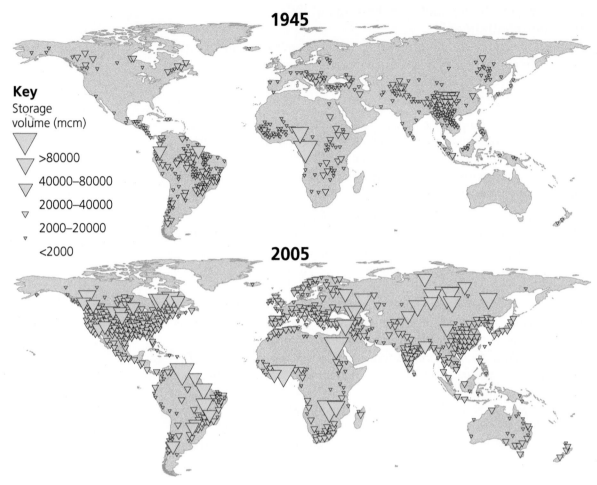

1945

Key
Storage
volume (mcm)

⬇ >80000

⬇ 40000–80000

▽ 20000–40000

▽ 2000–20000

▿ <2000

2005

▲ **Figure 3.20** The number of large dams worldwide, 1945 and 2005

Dam construction can impact significantly on the regime of a river (see page 82). The Colorado River had a complex regime prior to the construction of the Glen Canyon Dam in 1966 and the Hoover Dam in 1935.

● Originally, the river had a mostly very high flow between April and September. During the summer months, snowmelt in the Rocky Mountains and Winter River Mountains caused a significant rise in water level along the entire river. In the most extreme years of the early twentieth century, the Colorado's discharge was thirteen times higher in midsummer than in winter. Multiple peak flows were recorded from April onwards, each corresponding with snowmelt arriving from a different tributary river somewhere in the Colorado's enormous drainage basin (which covers seven US states). Mountain snowpack begins to melt in different upper regions of the catchment at slightly different times.

- Although river flow used to be generally low from September onwards, violent autumn storms over the Colorado Plateau contributed to another high, although shorter-lived, peak discharge in the autumn months.
- But the construction of large dams across the Colorado River has smoothed out the naturally occurring flood peaks and periods of low flows; the Hoover Dam now stores the equivalent of two years' river flow in Lake Mead. The mean annual flood at the gauging station at Lee's Ferry has been reduced from 2264 cubic metres per second to 764 cubic metres per second (according to data collected in the 1980s).

CONTEMPORARY CASE STUDY: WATER SUPPLY ISSUES IN YEMEN AND THE UAE

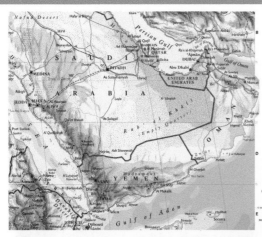

▲ **Figure 3.21** The physical geography of Yemen and the United Arab Emirates. ©Philip's

	Yemen	United Arab Emirates
Population	28 million	6 million
PPP ($)	2400	68,100
Crude birth rate (‰)	6.0	15.1
Crude death rate (‰)	6.0	1.9
Growth rate (%)	2.28	2.37
Life expectancy (years)	65.9	77.7
Infant mortality rate (‰)	46	10
Access to water (%) – urban	72	99.6
Access to water (%) – rural	46	199
Literacy (%)	70	93.8
Total fertility rate	3.63	2.32
Contribution to GDP (%) – agriculture	21.8	0.8
Contribution to GDP (%) – industry	9.8	39.5
Contribution to GDP (%) – services	68.4	40.1
Population below the poverty line (%)	54	19.5

▲ **Table 3.10** Factfile for the Yemen and the UAE

Source: *CIA World Factbook*, 2017

(NB The figures for UAE's contribution do not add up to 100%.)

Yemen's water challenge

One of Yemen's main challenges is severe water scarcity, especially in the highlands. Indeed, Yemen could become the first nation to run out of water, in particular its capital city, Sana'a. Another challenge is the high level of poverty, making it difficult to provide reliable safe water. Access to water supply is as low or even lower than in many sub-Saharan African countries. In addition, the ability of institutions to plan, build, operate and maintain infrastructure remains limited. Finally, the ongoing civil war since 2015 and military attacks by Saudi Arabia make it even more difficult to improve or even maintain existing levels of service.

	J	F	M	A	M	J	J	A	S	O	N	D
Temp. (°C)	24	25	27	28	30	32	32	31	31	28	27	26
Ppt. (mm)	5	2	5	2	2	2	5	3	2	2	2	5

▲ **Table 3.11** Climate data for Aden, Yemen

Enhanced water supplies in the UAE

▲ **Figure 3.22** A desalination plant in the UAE

	J	F	M	A	M	J	J	A	S	O	N	D
Temp. (°C)	21	21	25	30	33	36	38	38	35	32	25	23
Ppt. (mm)	19	25	22	7	0	0	0	0	0	1	3	16

▲ **Table 3.12** Climate data for Dubai, UAE

In contrast, on the opposite northern side of the Arabian Peninsula, the UAE is actively promoting the adoption of innovative technologies to reduce water wastage, protect the interests of water consumers and protect the environment. These pivotal actions enable the UAE to ensure a water-secure and sustainable future.

- The UAE is one of the top ten water-scarce countries in the world due to its hyper-arid climate – less than 100 mm/year rainfall.
- This country consumes more water than double the global national average.
- The UAE ranks as the world's second largest desalination producer (accounting for 14 per cent of the world's desalinated water). The Fujairah power and desalination plant, producing 455,000 cubic metres of freshwater per day, is the largest desalination hybrid plant in the world.
- Masdar City in Abu Dhabi (part of the UAE) recently launched a pilot programme for advanced energy-efficient seawater desalination technologies suitable to be powered solely by renewable energy sources.

▲ **Table 3.13** Key facts about water in the UAE

 # ⑤ Evaluating the issue

▶ *To what extent can water security ever be guaranteed in water-scarce places?*

Possible contexts for exploring water security and scarcity

Water security is defined in the United Nations Sustainable Development Goals as having access to sufficient amounts of safe drinking water. It is influenced by many factors, including precipitation amount and type, evaporation rates, climate change, pollution, competing demands for water, political stability and water infrastructure (such as the size, age and quality of pipes). Water is a vital resource but its availability varies spatially and temporally, and is not guaranteed in all places at all times, nor for all people.

Water-scarce places exist at many different scales:

- *Large continental areas of desert spanning multiple countries.* The Sahara and Sahel are notable examples where political control over exotic rivers such as the Nile or Colorado has become a vital factor affecting the water security of individual states.
- *Entire countries that are for the most part arid.* The examples of Yemen and the United Arab Emirates (UAE) in the Middle East illustrate the complex situation regarding water

security. Both areas are deserts. They share similar physical conditions – high temperatures, low rainfall and a lengthy coastline. However, there is considerable variation in the water security of both countries. Indeed, in some parts of the Middle East, water is considered to be more valuable than gold – it has been termed 'blue gold'.

- *Particular arid regions and local places found within some countries.* This is true of the USA, China and India for instance. There is also a great surplus of water in other, less populated, parts of these countries, thereby creating the possibility of water transfers to water-scarce areas.

Evaluating the view that water security *cannot* be guaranteed in water-scarce places

A lack of water security leads to increased morbidity and mortality. Around 1.8 billion people globally use a source of drinking water that is faecally contaminated. Water scarcity affects more than 40 per cent of the global population and is projected to rise. More than 80 per cent of wastewater resulting from human activities is discharged into rivers or sea without any treatment, leading to pollution.

Turning to examples of arid countries where water scarcity is widespread, Yemen, as we have seen (page 92), is the poorest country, and most water-stressed, in the Middle East. It is predicted that it may be the first country to run out of water – or that its capital city, Sana'a, could be the first capital city to run dry. The average Yemeni person receives just 140 m³ of water per year – compared with 1000 m³ for the average person in the Middle East. This is partly due to population growth (the rate of natural increase is slightly above 2 per cent) and its naturally arid desert location. Global climate change is influencing Yemen's water security too. Average rainfall in Sana'a dropped from 240 mm between 1932 and 1968, to 180 mm between 1983 and 2000. The water table at Sana'a was 30 m below the surface in 1970 – by 2015 it was 1200 m below the surface. Global climate change is adding to Yemen's problems. A combination of lower rainfall and higher temperatures reduces water availability, while continued population growth creates more demand for water for food production and drinking water, among others.

However, water insecurity is also a product of political factors in Yemen. Since the start of the 2015 Yemeni civil war, conditions have deteriorated. Up to 80 per cent of the country's population struggles to access drinking water. Water infrastructure has been targeted by planes. In 2016, a major desalination plant in the western city of Mokha was destroyed by a Saudi Arabian bomb. A blockade of imports resulted in a lack of oil required to pump groundwater. This forced many Yemeni to search for meagre sources of polluted surface water. The use of contaminated water has resulted in an increase in disease occurrence. In 2017, Yemen experienced over 1900 fatalities due to an outbreak of cholera. Moreover, some 14,000 under-five year olds die each year due to malnutrition.

Options are limited. The cost of pumping desalinated water from the Red Sea, some 250 km over the mountains that rise nearly 300 m, would push the price of water to US$10/m³. The cost of water has already risen 50 per cent since the start of the civil war, and some families spend up to a third of their income on water. This has an impact on how much they can spend on food. Typically the poorest households spend up to 80 per cent of their income on food, hence an obvious conflict of resource allocation occurs.

Human factors such as persisting poverty and conflict, along with longer-term climatic changes (see Chapter 5, pages 126–55), make it hard to see how water security improvements in Yemen and certain other countries including South Sudan and Somalia can be achieved.

Interactions between water-scarce neighbouring countries

Another important geographical scenario to consider is the way different water-scarce countries and regions may be dependent on a shared water source such as an exotic river or transboundary aquifer. Geopolitical factors – and a failure of global governance – can result in some countries suffering water scarcity due to the excessive demands placed on shared water resources by others. For example:

- The Nubian Sandstone Aquifer System under the eastern Sahara Desert is the largest aquifer in the world. It is believed to hold some 150,000 km³ of water. It is a fossil aquifer and is not currently being replenished. In recent years, Libya has developed its Great Man-made River Project in which it has been extracting 2.4 million km³ of water annually for irrigation purposes. This is the world's largest irrigation project.
- India, Nepal, Bhutan and Pakistan are involved in a huge 'water grab' in the Himalayas, as they seek new sources of electricity to power their economies. Altogether, the countries have plans for more than 400 hydro-electric dams which could provide more than 160,000 MW of

electricity. In addition, China has plans for around 100 dams to generate a similar amount of power from major rivers rising in Tibet. A further 60 or more dams are being planned for the Mekong River which also rises in Tibet. Hence, other downstream users may have access to less water in the future.

- The Grand Ethiopian Renaissance Dam, currently being built, will have a capacity of 66 km^3. When the reservoir is flooded, a process that could take between five and fifteen years, downstream countries (Egypt and Sudan) will have reduced flows, which will have serious impacts on their economic growth (agriculture and tourism, for example). International rivers are extremely difficult to manage, as increased use by upstream countries will affect countries downstream, both in terms of quantity of supplies and, possibly, water quality.

Evaluating the view that water security *can* be guaranteed in water-scarce places

As we have seen, some arid states *have* made great steps towards tackling the risk of water scarcity. The UAE (see pages 91–3), a country in the same region as Yemen and having a very similar climate, has delivered water security for its citizens. This is largely due to economic and political stability, however, which not all states enjoy. The demand for water in the UAE has increased with population growth, increased standard of living, industrialisation and the growth of the tourism sector. Much of the wealth of the UAE has been generated by the sale of oil and its reinvestment into other sectors, such as tourism, global trade and finance.

The UAE is an applied example of the Ester Boserup thesis of population and resources, i.e. that people will develop solutions to solve problems, once they arise. In Boserup's view, although rising population pressure could lead to resource depletion, human ingenuity and

flexibility will ultimately find new methods of technology and/or organisation in order to increase society's resource base.

- For example, the UAE has used desalinisation technology to increase water resources. Moreover, much of this is now done using renewable solar energy sources. The long-term goal is to extend renewable energy desalination throughout the UAE and MENA (Middle East–North Africa) region.
- The UAE's flagship sustainable city, Masdar City, has developed SMART-consumption, resulting in water consumption being over 50 per cent less than in other UAE cities. This includes smart-water meters, movement sensors instead of traditional taps, and hyper-efficient water appliances (washing machines and showers). Masdar's planners aim to reduce water consumption from 250 l/day at present to 105 l/day.
- Elsewhere in the world, modified irrigation techniques, such as drip irrigation, have been successfully introduced as a 'technological fix' for water scarcity.

Delivering water security through political co-operation

Turning to the issue of shared water sources, it is not inevitable that countries need suffer water insecurity due to a lack of political co-operation and agreements. At a national scale, Singapore illustrates how imports of water can help to make up the country's water supply. Singapore provides affordable high-quality water through supplies such as rainwater harvesting, reservoirs, desalination and water imports from Malaysia. It has been importing water from Malaysia since 1927 and has a deal that extends until 2061. Although it wishes to be self-sufficient in water by 2062, it currently imports around 66 per cent of the water it needs. Similarly, Hong Kong imports around 80 per cent of its water from Guangdong, China.

Water transfers can be achieved internally, i.e. *within* countries. The South-to-North Water

Diversion Project in China is the largest transfer of water between river basins in history. Since 2014 much of the drinking water consumed in Beijing has travelled over 1430 km from the Danjiangkou reservoir in central China. It takes fifteen days to reach Beijing. Two-thirds of the city's tap water and a third of its total supply now comes from Danjiangkou. In India, inter-basin transfers move excess water from the Brahmaputra and Ganga regions to dry areas such as Rajasthan, Gujarat and Tamil Nadu.

However, not all water management schemes are successful in all that they do – the River Colorado has failed to reach the sea most years since the 1960s, with harmful impacts for river and estuarine ecosystems. Only during El Niño weather events (which bring unusually large amounts of snow and rain to the Colorado Rockies and drainage basin) does the Colorado now ever reach the sea. Nevertheless, relatively effective water-sharing agreements between different US states and (to a lesser extent) Mexico continue to deliver water security to different user groups who depend on the continued benefits of irrigation, power generation, domestic and industrial water supplies and recreation.

Arriving at an evidenced conclusion

There are quantitative (measured) and qualitative (experienced) aspects to water security across a variety of spatial and temporal scals. Some places lack sufficient amounts of water; and not all available water is safe. Poor water quality jeopardises human health over a large part of the Earth's land surface.

Even in the same world region, such as the Middle East, there are widely different experiences of water security. In the case of Yemen, a poor country, much of its water is needed for food supply. This country lacks the funds to develop desalination plants and pump water from underground aquifers, while civil war has destroyed parts of its water infrastructure, reducing the country's ability to utilise groundwater. In contrast, the UAE – where natural freshwater resources are severely limited in size, and there are increasing demands due to population growth and rising affluence – is a place where solutions for water scarcity have been found, albeit at great expense. The UAE shows that water security *can* be guaranteed provided there is sufficient economic and political stability.

In some areas, political co-operation can help deliver water security to multiple countries, states and/or different user groups. Internal transfers of water – for example, from southern China to northern China – can guarantee water security for places where water is naturally relatively scarce. However, the extent to which current strategies and 'fixes' are future-proof is debateable, especially in emerging economies where populations and average incomes are still rising. Ultimately, lifestyle changes may be necessary: for example, typical per capita water footprints would be reduced if people ate less meat and repaired (rather than replaced) manufactured goods. In addition, further reductions in population growth might ease pressures on water resources: whether world population growth eventually peaks at around 9 billion or 11 billion is probably the most important factor affecting future water scarcity both globally and locally.

🔑 KEY TERMS

Exotic river This starts out from a humid region and flows into a dry region. A river flowing through a desert can be considered exotic due to its existence in an otherwise arid region.

Transboundary aquifer A store of groundwater that extends over a number of countries, e.g. the Nubian Sandstone Aquifer System which stretches across the eastern Sahara, under Egypt, north-west Sudan, north-east Chad and eastern Libya.

Chapter summary

✓ The water budgets of different places vary greatly because of vastly differing temperature and rainfall patterns.

✓ Some places, such as those with a monsoonal climate, enjoy a water surplus at some times of the year and a deficit at other times, creating challenges for water management.

✓ Billions of people live in regions that experience some degree of water scarcity attributable to aridity or recurring drought; many hundreds of millions lack access to adequate safe water for much of the year.

✓ Water availability is also affected by human factors, including the virtual water transfers associated with trade, water transfer schemes, dam building and land grabs.

✓ There are many technologies that can help improve water supplies and the efficiency with which it is used, including desalination and efficient forms of irrigation such as drip irrigation. However, these can be costly to introduce and not all developing countries have the means to afford them.

✓ In large river basins, effective water management may require a high level of international co-operation and agreements.

Refresher questions

1 What is meant by the following geographical terms: water budget; soil moisture deficit; soil moisture recharge; steady state equilibrium?

2 Using examples, outline the difference between (i) physical water scarcity and economic water scarcity, (ii) improved and unimproved water sources, (iii) drought and aridity.

3 Using examples, explain the water footprint concept.

4 Briefly outline the main features of the changing global pattern of water use.

5 Using examples, explain why water availability in a country may vary (i) seasonally, (ii) between urban and rural areas.

6 Using examples, explain what is meant by the following geographical terms: land grab; virtual water?

7 Explain why some irrigation methods are better than others in terms of their efficiency of use of hydrological resources.

8 Explain the challenges and opportunities associated with large-scale desalinisation as a potential 'technological fix' for water insecurity in contrasting places you have studied.

9 Explain how dam construction can (i) modify water cycle flow and storage patterns in a drainage basin, (ii) become a source of geopolitical tension.

Discussion activities

1 In small groups, discuss the relative importance of physical and human factors as a cause of global water insecurity. In your view, are human or physical factors ultimately most important?

2 In pairs, devise a long-term research programme lasting several years to create a water budget model

(such as Figure 3.1) for a place that has never been studied before. How would you collect the data that is needed? How many years would the study need to run for before researchers were satisfied that their data was representative of a typical year?

3 As a whole class activity, discuss the size of your own water footprint(s), taking into account drinking, bathing, cleaning.

4   In small groups, discuss your consumption of virtual or embedded water, focusing on fruit and vegetables imported from other countries and the place of origin for clothing and other manufactured goods you own. Are you aware of any products you consume that have originated in countries where water is scarce?

5   As a whole class activity, discuss the view that water scarcity will eventually become a thing of the past on account of technological progress.

FIELDWORK FOCUS

The topic of access to (safe) water can be used as the basis for a number of different A-level individual investigation themes.

A   *Investigate the players who have power over water supplies in your local area.* Secondary data can be collected by monitoring your local newspaper (and their website) to investigate water issues in your area, along with the Environment Agency website. For primary data, (i) identify the different organisations involved in managing water, and (ii) arrange to carry out an in-depth interview with one or more managers. You could conduct a questionnaire in the area to find out more about people's perception of local water supply issues and how they are being managed.

B   *Investigate the global flows of virtual water which link your school (or another local institution) with other places (and water cycles) around the world.* You could arrange for an interview with a school chef or administrator in order to find out where the food the school uses is sourced from, which will most likely include local and global production networks. Secondary research will help you find out how much water is embedded in certain items such as bananas, rice or potatoes. By collecting data about the volumes of food purchased by your school, you can begin to estimate the total virtual water demand (and use world maps to show where the flows come from). You might also carry out a questionnaire among students and staff to investigate their awareness of the water footprint concept.

C   *Investigate soil moisture status (as a proxy for water availability) in your local area.* Primary data can include soil moisture content – you will need to borrow or buy specialist equipment for this task, however, such as a soil moisture gauge (these can be purchased easily online). Monitoring will ideally need to be done over an extended period of time, and should be carried out in conjunction with readings for precipitation and temperature. Ideally, the data will be collected over several months in order to include a good balance of wet days and dry days, and possibly changes in temperature. It may be possible to get figures for actual and potential evapotranspiration by logging into the Meteorological Office Rainfall and Evaporation Calculation System (MORECS).

Further reading

Ward, C. (2015) *The Water Crisis in Yemen: Managing Extreme Water Scarcity in the Middle East*, I.B. Taurus

United Nations Sustainable Development Goals – water and sanitation, www.un.org/sustainabledevelopment/water-and-sanitation

www.unwater.org/publication_categories/sdg-6-synthesis-report-2018-on-water-and-sanitation

The World's Water, http://worldwater.org

World Bank, www.worldbank.org/en/topic/water

World Health Organisation, www.who.int/topics/water/en

Carbon cycle dynamics

A balanced carbon cycle is important in maintaining the health of the planet. Physical processes operating at a range of spatial scales control the movement of carbon between stores on land, the oceans and the atmosphere. This chapter:

- explores the global carbon cycle and its stores
- analyses carbon flows and processes
- investigates carbon cycling at the local scale
- assesses the spatial variability of system flows.

KEY CONCEPTS

Biogeochemical cycle Pathway by which a chemical element moves between non-living (abiotic) and living (abiotic) components of the Earth.

Mass balance At a global scale, this means that the total amount of carbon is conserved over time.

Scale Places can be identified at a variety of geographic scales, from local territories to the national or state level. Global-scale interactions occur at a planetary level.

System An assemblage of parts and the relationships between them, which together constitute an entity or whole. The systems approach helps us visualise complex sets of interactions.

1 The global carbon cycle and its stores

▶ *Where is carbon stored and how do these storages vary in size and extent?*

 **KEY TERM**

Photosynthesis The process whereby plants use sunlight to combine carbon dioxide (CO_2) from the air with water to make complex molecules.

In a closed system, there is no transfer of energy or matter across the external boundaries of the system (see Chapter 1, Figure 1.8a). The carbon cycle is often considered to be a closed system, as the supply of carbon remains within the Earth system. However, **photosynthesis** is driven by the input of solar energy, and so in reality there is the movement of energy (if not matter) across the boundary of the Earth. Matter cycles between abiotic and biotic environments. The abiotic environment includes the atmosphere, lithosphere, hydrosphere and cryosphere. The biotic component is the biosphere. The supply of nutrients in an ecosystem is therefore finite and limited, whereas in contrast there is a continuous, if variable, supply of energy in the form of sunlight.

All chemical elements circulate between living things and the abiotic environment. The process of exchange or 'flux' (back and forth movements) of materials is a continuous one. These movements have a biological component and a geochemical component and so consequently are known as biogeochemical cycles. Continued availability of carbon in ecosystems depends on carbon cycling.

Carbon-based life

About 96 per cent of living matter consists of just four elements: carbon, hydrogen, oxygen and nitrogen. Life on Earth is described as 'carbon-based' because the main structural element in biomolecules is carbon, making up around 18.5 per cent of human total body mass. Carbon atoms are able to react with each other, and other elements, to form extremely stable molecules. At least 2.5 million organic (carbon-based) compounds exist, the largest group of which are the carbohydrates, which include sugars, cellulose, starch and glycogen.

Lipids and proteins also contain carbon. Proteins make up about two-thirds of the total dry mass of a cell, and lipids are important molecules for energy storage and insulation. The atomic elements that molecules are made from, such as carbon, cycle within ecosystems, entering the biotic system and leaving it to return to the abiotic environment.

Stores of carbon

The places where carbon resides in the global system, such as in living organisms, are called stores. As well as being contained in organic molecules, carbon is also present in carbon dioxide (CO_2) and methane (CH_4), another atmospheric gas containing carbon – both are part of the air we breathe. Carbon exists in a dissolved form in water, and is present in limestone, fossil fuels, ocean sediments and soils.

The biogeochemical carbon cycle consists of carbon stores of different sizes (terrestrial, oceans and atmosphere). The amounts of carbon held in different stores are shown in Table 4.1. The biggest store of carbon is in limestone, produced from shells and reef-building coral. The residence time of carbon is the average length of time it remains in any carbon store: this ranges from long term (millions/thousands of years) to short term (tens/hundreds of years).

Store	Carbon in store (gigatonnes)
Sedimentary (carbonate) rocks and deep ocean sediments	100,000,000
Ocean water (biomass and dissolved CO_2)	38,700
Sea floor sediments	6000
Fossil fuels (coal, oil, gas)	4130
Soils and peat	2300
Atmosphere (gaseous CO_2)	600
The biosphere (forests and grasslands)	560

▲ **Table 4.1** Global carbon stores (figures are naturally occurring volumes and do not take into account human activity)

Figure 4.1 shows the global carbon cycle in diagrammatic form. Carbon flows through these stores are driven by a range of natural processes, which are explored in the next section (pages 102–10).

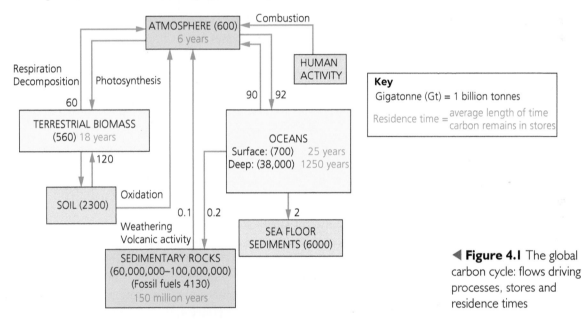

◀ **Figure 4.1** The global carbon cycle: flows driving processes, stores and residence times

Figure 4.1 shows that the majority of global carbon is locked in terrestrial stores. Most of the terrestrial stores of the Earth's carbon are geological, resulting from the formation of sedimentary carbonate rocks (limestone) in the oceans and biologically derived carbon in shale, coal and other rocks. These stores take part in the long-term geological cycle.

Carbon stores in different biomes

When human activity is taken into account, the total amount of carbon stored in the terrestrial biosphere (soil and biomass) is estimated to be approximately 3000 GtC. This storage is spread unevenly among the different terrestrial biomes, as Figure 4.2 shows. Forests are significant

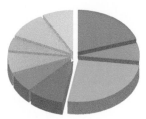

Key
- ■ Tropical forest 20%
- ■ Temperate forest 7%
- ■ Boreal forest 26%
- ■ Agriculture 9%
- ■ Wetlands 7%
- □ Tundra 8%
- ■ Desert 5%
- ■ Temperate grassland 10%
- □ Tropical savanna 8%

▲ **Figure 4.2** The carbon biomass storage contribution of different world biomes

 **KEY TERM**

Gigatonne A gigatonne (Gt) amounts to one billion tonnes (or one trillion kilogrammes). A gigatonne of carbon dioxide equivalent (GtC) is the unit used by the United Nations climate change panel, the Intergovernmental Panel on Climate Change (IPCC), to measure the amount of carbon in various stores.

carbon stores; they make up more than half of all biomass storage, where carbon is stored in the biomass of trees (above ground biomass). A thick litter layer on the forest floor can also store significant amount (part of the soil storage).

- Boreal (coniferous) forest stores more carbon than tropical rainforest globally because it is distributed across a greater area, including much of northern Russia and North America.
- The biomass of tropical rainforest (the weight of organic matter per unit area) is greater than that of boreal forest but it does not cover as great a land area.
- Carbon is an essential plant macronutrient and makes up approximately 44 per cent of the dry weight of plant biomass. This is because it is a major component of all organic molecules (see above).

② Carbon flows and processes

▶ *What biological and geological processes move carbon between stores, and how do these vary both temporally and quantitatively?*

Physical and biological processes control the movement of carbon between stores on land, the oceans and the atmosphere. Movements or transfers between the stores of the carbon cycle are called flows. For example, volcanic activity adds 0.1 gigatonnes of carbon to the atmosphere every year. The rate of flow (measured as units of mass per unit time) is sometimes called a flux. In systems diagrams, flows into or out of particular stores are called inputs and outputs.

Processes are the physical and biological mechanisms that drive the inputs, outputs and flows of the cycle. (Chapter 2 explored how processes such as evaporation and infiltration drive the flows of the water cycle.) Carbon cycle processes include: photosynthesis, respiration, decomposition, combustion (including natural and fossil fuel use), natural sequestration in oceans, vegetation, sediments and weathering (Table 4.2).

	Processes by which carbon is circulated	Operational timescale
Combustion	• Releases CO_2 into the atmosphere. • Since the start of the industrial revolution in Europe, CO_2 has been released at an increasing rate by the combustion of biomass and fossilised organic matter.	• seconds
Diffusion	• Carbon dioxide diffuses from the atmosphere or water into autotrophs (plants and other photosynthetic organisms).	• seconds

Processes by which carbon is circulated		Operational timescale
Photosynthesis	• Autotrophs convert CO_2 from the atmosphere into carbohydrates and other organic compounds. • Aquatic plants use dissolved CO_2 and hydrogen carbonate ions (as HCO_3^-) from the water in the same way.	• seconds
Respiration	• CO_2 is produced as a product of metabolism and diffuses out into the atmosphere or water.	• seconds
Methane formation	• Organic matter held under anaerobic conditions (such as in waterlogged soil or in the mud of deep ponds) is metabolised by methane-producing bacteria (methanogenic archaeans). • Methane accumulates in the ground in porous rocks or underwater, but may diffuse into the atmosphere. • In air and light, methane (CH_4) is oxidised to CO_2 and water.	• minutes
Feeding/consumption	• Animals feed on plants to obtain glucose and other nutrients. Biomass is passed from one animal to the next in a food chain.	• minutes/hours
Decay/decomposition	• Dead organic matter is decomposed to CO_2, water, ammonia and mineral ions by microorganisms. • The CO_2 produced diffuses out into the atmosphere or dissolves in water (as HCO_3^- ions).	• hours/days
Volcanic eruptions	• Sedimentary rocks move down into the mantle (subduction). • When volcanoes erupt, CO_2 is released into the atmosphere.	• hours/days
Peat formation	• In acidic and anaerobic conditions found in waterlogged soils, dead organic matter is not fully decomposed but accumulates as peat. • Peat decays slowly when exposed to oxygen, releasing CO_2 into the atmosphere.	• centuries
Fossilisation	• Partially decomposed organic matter from past geological eras was converted either into coal or into oil and gas that accumulate in porous rocks. • Peat in past geological eras was converted to coal, oil or gas.	• millennia
Shell and coral formation	• Many organisms combine HCO_3^- with calcium ions to form calcium carbonate shells and coral skeleton (animals such as reef-building corals and mollusca such as shellfish). • Shells and reef-building coral can become fossilised and form sedimentary rocks (chalk and limestone) over long periods of geological time.	• millennia

▲ **Table 4.2** How carbon is circulated between stores, and one view of the different timescales over which processes operate

Figure 4.1 (page 101) shows carbon fluxes due to processes in the carbon cycle. In the carbon cycle, processes transfer carbon from one store to another. Global carbon fluxes are very large, and are measured in gigatonnes (Gt), where $1\ Gt = 1 \times 10^{15}$ grams.

Proportional flow diagrams can be used to show the relative proportions of different flows (Figure 4.3).

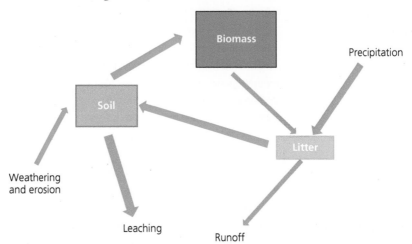

▲ **Figure 4.3** Diagram showing a nutrient model for a rainforest ecosystem. The size of the boxes represents the amount of nutrients stored, and the width of the arrows the size of the nutrient flows

Biological processes of the carbon cycle

Photosynthesis

Green plants use the energy of sunlight to produce sugars from the inorganic raw materials carbon dioxide and water, by a process called photosynthesis. The waste product is oxygen. Photosynthesis in global forests cycles approximately one-twelfth of atmospheric carbon dioxide every year, accounting for 50 per cent of terrestrial photosynthesis.

Green plants and algae in the oceans also maintain the composition of the atmosphere today. For example, the quantity of carbon dioxide removed by plants in photosynthesis each day is almost equal to that added to the air from respiration (see below) and from the burning of fossil fuels. Photosynthesis is also the only natural process that releases oxygen into the atmosphere. All the oxygen present in the air (about 21 per cent) is a waste product of photosynthesis. This is the source of oxygen for aerobic respiration.

Respiration

The steps of aerobic respiration can be summarised by a single equation:

glucose + oxygen → carbon dioxide + water + ENERGY

$$C_6H_{12}O_6 + 6O_2 \rightarrow 6CO_2 + 6H_2O + ENERGY$$

This equation is a balance sheet of the inputs (raw materials) and the outputs (products). The energy released by respiration is used to support processes such as active transport, movement, synthesis of molecules, responding to the environment (sensitivity), reproduction and excretion.

Consumption

Animals, in contrast to plants, obtain nutrients as complex organic molecules of food which they digest, absorb and assimilate into their own cells and tissues. Animals are heterotrophs, which means they cannot make their own sugars in the same way that autotrophs can, and must obtain these from other organisms by feeding on them.

Decomposition

Recycling of nutrients is essential for the survival of living organisms, because the available resources of many elements are limited. When organisms die, their bodies are broken down to simpler substances (for example, CO_2, H_2O, NH_3 and various ions) and nutrients are released. Bacteria and fungi, feeding on dead organisms or waste material, respire and release carbon back into the atmosphere as carbon dioxide. The elements that are released may become part of the soil solution, and some may react with chemicals of soil or rock particles, before becoming part of living things again through reabsorption by plants.

> Decomposition is the breakdown of plant material to simpler organic molecules which are then the fuel for respiration. This breakdown is facilitated by enzymes released by decomposing microbes (fungi and bacteria). Physical breakdown is also promoted by some higher organisms (invertebrates).

Geological processes of the carbon cycle

Most global carbon is geological, resulting from the formation of sedimentary carbonate rocks (limestone) in the oceans and biologically derived carbon in shale, coal and other rocks. These sources of carbon are locked in terrestrial stores as part of the long-term geological cycle. Carbon cycles between the land and ocean are sometimes called the 'slow carbon cycle'. The cycling of carbon between bedrock stores on the land and the oceans occurs through processes of weathering, erosion and deposition over very long timescales (millions of years), and at a continental scale (Figure 4.4). Geological processes release carbon into the atmosphere through volcanic out-gassing at ocean ridges/subduction zones and chemical weathering of rocks.

Water is the key medium for chemical weathering. Acidic water helps to break down rocks such as chalk and limestone. Most weathering therefore takes place above the water table, since weathered material accumulates in the water and saturates it. One type of chemical weathering, carbonation-solution, releases carbon dioxide from rock stores. It occurs on rocks with calcium carbonate, such as chalk and limestone. Rainfall combines with dissolved carbon dioxide or organic acid to form a weak carbonic acid:

$$CO_2 + H_2O \leftrightarrow H_2CO_3 \text{ (carbonic acid)}$$

Calcium carbonate (calcite) reacts with an acid water and forms calcium bicarbonate (also termed 'calcium hydrogen carbonate'), which is soluble and removed by percolating water:

$$CaCO_3 + H_2CO_3 \rightarrow Ca(HCO_3)_2$$
calcite + carbonic acid $\rightarrow$ calcium bicarbonate

Globally, some 0.3 billion tonnes of carbon are transferred from rocks to the atmosphere and oceans each year by chemical weathering.

Calcium bicarbonate can be transferred by overland flow, throughflow or groundwater flow into rivers where it becomes part of the solute load (dissolved material carried in solution, an important river transport mechanism). Water moves slowly downwards from the soil into the bedrock (percolation – see Chapter 2, pages 37 and 44): in rocks such as carboniferous limestone and chalk this process may be quite fast, locally.

● Over time, large amounts of carbon have been removed in solution from limestone areas of the UK, such as the cliffs at Llandudno.
● Carbonation weathering takes place when rainwater collects in pools on the surface of exposed rock, for example the limestone pavement at Malham, Yorkshire.
● In chalk regions such as England's South Downs, the slow movement of groundwater dissolves the rock it is transmitted through, and eventually transfers calcium bicarbonate in solution into river systems and ultimately the ocean.

On a local scale, underlying geology affects percentage carbon content of rivers. For example, if the drainage basin of a river is an area of chalk grassland, percentage carbon solute content of water will be higher than a river that has a drainage basin of an upland area of land underlain by impermeable granite.

Once in the ocean, carbonate is used by marine organisms to create shells. When they die, these organisms' carbonate shells are deposited as carbonate-rich sediment on the ocean floor where they are eventually lithified (turned into rock). This part of the carbon cycle can lock up carbon for millions of years. It is estimated that the oceanic sedimentary layer may store up to 100 million GtC.

Figure 4.4 shows how huge volumes of stored carbon are constantly on the move (albeit at extremely slow rates of movement) through the geological process of tectonic plate movement.

Some carbon is eventually returned to the atmosphere by volcanism, as CO_2 is released from melted rocks when subduction occurs at plate boundaries.

▶ **Figure 4.4** The slow geological (land–ocean) carbon cycle. Less CO_2 is released by chemical weathering than is taken up by carbonate deposition so that the weathering and ocean deposition part of the cycle is a net drawdown of carbon from the atmosphere. At long timescales this is balanced by volcanic release.

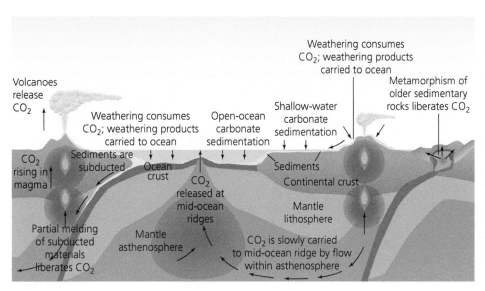

CONTEMPORARY CASE STUDY: CAMEROON'S LAKE NYOS GAS BURST

An example of carbon dioxide released by volcanic activity occurred in 1986, when a large volume of carbon dioxide was released by Lake Nyos, a crater lake in northern Cameroon. The source of carbon dioxide was a basaltic chamber of magma, deep beneath Cameroon, which had been leaking into and accumulating in Lake Nyos for some time. Due to its depth, water in the lake became stratified into layers of warmer water near the surface and colder denser water near the bottom of the lake. The cold dense water absorbed the carbon dioxide, which was then held down by the weight of the overlying waters.

A disaster occurred after the water at the bottom of the lake was disturbed, possibly due to a deep volcanic eruption, an earthquake, and a change in water temperature or a climatic event. The released gas swept down into neighbouring valleys for a distance of up to 25 kilometres (Figure 4.5). Carbon dioxide is denser and heavier than oxygen, and so travels along the ground, causing oxygen-deprivation (asphyxiation) in people and animals. Some 1700 people were suffocated, 3000 cattle died and all other animal life in the area was killed.

Following the tragic Nyos event, volcanic lake research concerning volcanic lakes grew globally. In 1987, the International Conference on the Lake Nyos Gas Disaster, held in Cameroon's capital, Yaoundé, established the International Working Group on Crater Lakes (IWGCL). Later, in 1993, the International Association of Volcanology and Chemistry of the Earth's Interior (IAVCEI) recognised IWGCL as a commission, renaming it the Commission on Volcanic Lakes (CVL).

In March 2016, 30 years after the gas burst, the IAVCEI-CVL met in Yaoundé for a

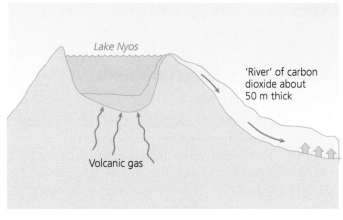

▲ **Figure 4.5** Lake Nyos, Cameroon

three-yearly workshop, to discuss progress made since the disaster. Many local students and 65 professional participants from 11 countries attended the workshop. The meeting recognised the role of the international community in the decades-long research on volcanic lakes, in particular the role of the Japan International Co-operation Agency, which has resulted in several Cameroonian experts returning home having obtained PhDs in Japan, to head up research on their own lakes. Workshop attendees recognised that research efforts stand as an outstanding example of international collaboration and capacity building – both of which may be regarded as important facets of effective global governance (see also pages 156–7).

Carbon cycle flows from the atmosphere to the ocean

The oceans take up carbon dioxide by two carbon cycle 'pump' mechanisms referred to as (1) the physical pump, and (2) the biological pump (Figure 4.6).

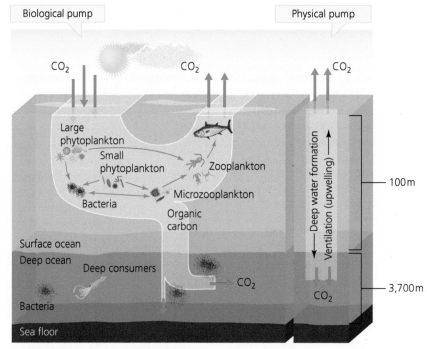

▲ **Figure 4.6** Oceanic carbon pumps

1 The physical (inorganic) pump involves the movement of carbon dioxide from the atmosphere to the ocean by a process called diffusion. CO_2 dissolved in the surface of the ocean can be transferred to the deep ocean in areas where cold dense surface waters sink. This downwelling carries carbon molecules to great depths where they may remain for centuries. The level of CO_2 diffusion also determines the acidity of the oceans.

2 The biological (organic) pump is driven by ocean phytoplankton absorbing carbon dioxide through photosynthesis. These organisms form the bottom of the marine food web, and they live in the ocean's

surface layer (euphotic zone) where sunlight penetrates. Phytoplankton are consumed by other marine organisms and carbon is subsequently transferred along food chains by fish and larger sea animals as they consume one another. Organic carbon may eventually be transferred to the deep ocean when dead organisms sink towards the ocean floor. Other carbon is absorbed in calcite shell formation (see page 102).

In summary, both very fast (photosynthesis) and slower (downwelling) processes are involved in the flow of carbon into and out of ocean storage.

The concept of mass balance

The concept of mass balance can be applied to the carbon cycle: this means that, at a global scale, the total amount of carbon is conserved over time (although long-term changes may occur where it is stored as a result of climate change and sedimentary processes). A balanced carbon cycle is important in sustaining other Earth systems but is increasingly altered by human activities (see pages 140–1).

ANALYSIS AND INTERPRETATION

Figure 4.7 shows the main components of the carbon cycle. The figure indicates the main stores and the principal fluxes of carbon for the natural carbon cycle, prior to anthropogenic disturbance (i.e. 'pre-carbon disruption' era). There are four main carbon stores: the geological, the oceanic, the terrestrial and the atmospheric (in decreasing order of size).

The thick arrows in Figure 4.7 indicate the most important fluxes for the current balance of CO_2 in the atmosphere:

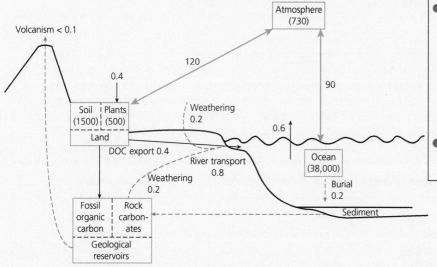

- arrow from atmosphere to land: gross primary production (i.e. absorption of carbon from the atmosphere through photosynthesis); arrow from land to atmosphere: respiration (release of carbon to the atmosphere)
- arrows to and from ocean and atmosphere: physical sea-air exchange.

▲ **Figure 4.7** Main components of the natural carbon cycle. Thick arrows signify photosynthesis and respiration by the biosphere and sea-air exchange. Thin arrows signify natural fluxes which are important over longer timescales. Dashed lines represent fluxes of carbon as calcium carbonate. The units for all fluxes are PgC yr^{-1} (petagrams of carbon per year); the units for all stores are PgC.

Under 'natural' conditions these two main carbon transfer pathways between the atmosphere and the land or the oceans are in approximate balance over an annual cycle:

- They represent a total exchange of 210 PgC yr⁻¹

- The larger share (120 PgC yr⁻¹) is taken by the land.

This annual exchange is more than 25 times the total amount of carbon annually released into the atmosphere through human activities. Forests are responsible for about half of total terrestrial photosynthesis, and hence for roughly 60 PgC yr⁻¹.

Other fluxes in the system include dissolved inorganic carbon (DIC) derived from the weathering of carbonates in rocks, and outflow of dissolved organic carbon (DOC) through rivers to the sea.

(a) Figure 4.7 shows inputs from photosynthesis and outputs from respiration, and inputs and outputs from the physical sea-air exchange.

  i Describe additional fluxes shown in the figure and state quantities of carbon involved.

  ii Explain the role of each additional flux in the carbon cycle.

GUIDANCE

This question asks you to interpret a systems diagram to assess which additional flows are shown. The quantities are indicated by both the width of the arrows and the numbers given. The role in the carbon cycle can be interpreted from the direction of the flows and the stores that they link, as well as applying your knowledge of the carbon cycle at both short and long timescales.

(b) The diagram shows the balance between stores and fluxes in the 'pre-carbon disruption' era. Suggest how the expansion of human populations across the globe has affected the carbon cycle.

GUIDANCE

This is a broad question which allows different aspects of system dynamics to be explored. One approach might be to suggest how particular flows and stores have changed in size on account of particular activities. Additionally, we might consider changes in the global pattern of carbon storage, or changes in the speed of fluxes (transfers in agricultural systems could be faster than for some naturally occurring plant communities). Finally, a good answer might apply the concept of equilibrium (by suggesting the extent to which changes caused by humans have disturbed overall system balance at global and local scales).

(c) Suggest which aspects of the natural carbon cycle will not have been affected by human expansion.

GUIDANCE

This links to the previous question, and requires an appreciation of the different timescales at which the carbon cycle operates. Longer timescale processes, for example, exert an important influence on atmospheric CO_2 concentrations on geological timescales (millions of years), but have had little influence at the timescale corresponding to human expansion (100,000 years).

③ Carbon cycling at the local scale

▶ *How do the processes of photosynthesis and respiration interact to affect carbon flows and storages at a local scale, and what role does climate play?*

Fast carbon cycle flows between the land and the atmosphere

At a short-term and local scale, several important processes are involved in carbon sequestration and release: photosynthesis, respiration, decomposition and fossil fuel combustion. These are sometimes referred to as 'fast carbon cycle' processes (Figure 4.8).

Carbon sequestration is the natural capture and storage of carbon dioxide (CO_2) from the atmosphere by physical or biological processes such as photosynthesis.

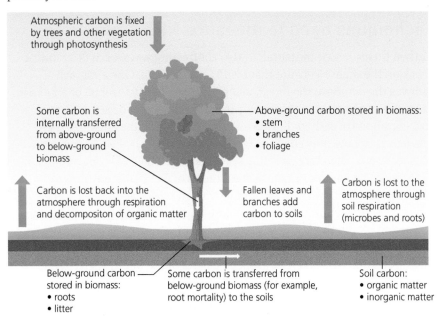

Atmospheric carbon is fixed by trees and other vegetation through photosynthesis

Some carbon is internally transferred from above-ground to below-ground biomass

Above-ground carbon stored in biomass:
• stem
• branches
• foliage

Carbon is lost back into the atmosphere through respiration and decompositon of organic matter

Fallen leaves and branches add carbon to soils

Carbon is lost to the atmosphere through soil respiration (microbes and roots)

Below-ground carbon stored in biomass:
• roots
• litter

Some carbon is transferred from below-ground biomass (for example, root mortality) to the soils

Soil carbon:
• organic matter
• inorganic matter

▲ **Figure 4.8** Carbon sequestration and losses as a result of transfers between the atmosphere and land (including vegetation)

Ecosystem productivity

A very small amount of the energy available to producers ends up in food chains. Much of the light energy is reflected, transmitted through leaves or is the wrong wavelength for photosynthesis. Of the energy that is stored in producers (around 1 per cent of that available from the Sun), a decreasing

KEY TERMS

Autotrophic Relating to organisms that can make their own food via photosynthesis. All plants, some protoctistans and bacteria are autotrophic.

Gross primary productivity (GPP) The total gain in energy or biomass fixed by photosynthesis in green plants per unit area per unit time.

Net primary productivity (NPP) The gain by producers in energy or biomass, per unit area per unit time, remaining after allowing for respiratory losses from producers (R). This biomass is available to consumers in an ecosystem.

amount is available as it passes through the food chain. The products of this **autotrophic** nutrition are then eaten by consumers.

The amount of glucose produced by the plant in a specific area in a specific period of time is known as **gross primary production** (or **GPP**). This equates to the rate of photosynthesis. Much of this energy is used in respiration and ultimately lost as heat from the producers. The remaining energy is stored in new biomass: this is known as **net primary production** (or **NPP**).

NPP can be calculated as follows:

$$NPP = GPP - R$$

where R = respiratory loss.

NPP is the rate at which plants accumulate new biomass. It represents the actual store of energy contained in potential food for consumers. NPP is easier to calculate than GPP as biomass is simpler to measure than the amount of energy fixed into glucose. Between 30 and 50 per cent of gross primary production (GPP), i.e. the total amount of carbon absorbed by vegetation, is used to support the metabolic processes of plants, and is released back into the atmosphere when plants respire.

Techniques used to measure NPP

Carbon fluxes can be measured in two contrasting ways, each providing a check on the other to ensure that results are accurate. One approach samples the air above the forest, using moving *eddies* (currents) of air above the forest canopy to simultaneously measure the differences between concentrations of carbon dioxide and the direction of the eddy (i.e. their *covariance*, or how they change together), and so is called **eddy covariance**. This technique therefore uses fine scale measurements of air movement and CO_2 concentration to estimate vertical fluxes of CO_2. This 'top-down' approach gives an overview of productivity and respiration rates of the forest. The other approach involves more traditional fieldwork, measuring vegetational growth, storage and rates of respiration on the forest floor – these '**biometric**', or 'bottom-up', measurements allow fine-scale measurements to be taken from specific parts of the forest, such as soil, stem and litter layer.

Results from both top-down and bottom-up approaches can be compared to establish a set of consistent estimates of productivity and **ecosystem respiration (Re)**: concordant results that are known to accurately represent carbon flows in the forest. Data from both techniques can be recorded both monthly and annually to ensure resolution at a range of temporal scales.

Eddy covariance techniques

Eddy covariance is the most accurate way to measure carbon uptake and loss from ecosystems. Measurements are taken from a tower within the sampling area, 25 metres above the forest floor. Sampling can be done using a Solent R2 3D sonic anemometer and an open-path infra-red gas analyser.

KEY TERMS

Eddy covariance A technique that measures wind speed and carbon dioxide concentrations to quantify fluxes to and from ecosystems.

Biometric measurements Direct measurements taken of different biotic components of the ecosystem, incorporating a variety of fieldwork techniques.

Ecosystem respiration (Re) Combined autotrophic and heterotrophic respiration, resulting in loss of carbon from the ecosystem to the atmosphere.

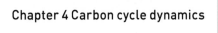

Updraughts and downdraughts of air are the result of turbulent eddies in the air, caused by temperature fluctuations. Recording changing speeds of updraughts and downdraughts allows scientists to calculate the net release uptake of carbon dioxide. The sonic anemometer measures wind speeds in three dimensions, by firing ultrasonic soundwaves in different directions. The infra-red gas analyser uses an infra-red light source to measure carbon dioxide concentrations. Each of the instruments can simultaneously collect thousands of sets of data per minute. Net fluxes can be estimated by calculating differences in carbon dioxide molecules moving upwards away from the ground and downwards towards it, within specific eddies of air.

In addition to measuring carbon dioxide concentrations, the tower also collects other data. A probe measures temperature and relative humidity, and other sensors measure incoming solar radiation and photosynthetically active radiation (PAR), the latter allowing measurements to be taken of how much light is available to plants for photosynthesis. This data can be used to correlate with estimations of **net ecosystem productivity (NEP)**.

> **KEY TERM**
>
> **Net ecosystem productivity (NEP)** The difference between GPP and fluxes due to ecosystem respiration (R_e). Positive NEP indicates net sequestration of carbon.

> Measurements taken by eddy covariance are expressed as megagrams of carbon (MgC), where 1 Mg = 1000 kilograms, or 10^6 grams.

Data from all eddy covariance studies contributes to an international database, Fluxnet, which contributes to global models of carbon fluxes.

Biometric techniques
Biometric techniques use ground-based ecological methods. Dendrometer bands (strips placed around the circumference of a tree and used to make sequential repeated measurements of tree growth) and Vernier-gauge callipers are used to measure to a precision of 0.5 mm. Scaling techniques, specific for each species of tree, are used to estimate the above-ground woody biomass of each tree. Growth in the circumference of the tree, and therefore increase in biomass, can be converted into estimates of productivity. The production of coarse woody debris (fallen dead trees and the remains of large branches on the ground in forests) can be estimated using mortality recorded during DBH measurements.

Leaf production is estimated by sampling leaf-fall in the autumn, using traps to catch falling leaves. Similar traps can be used to sample flower and fruit production later in the year, allowing the productivity of reproductive organs to be collected.

Chamber measurements
Below-ground root productivity is estimated using measurements of inputs to the soil, from root mortality and litter-derived material, and rates of respiration. Respiration rates from different parts of the forest can be measured using a portable infra-red gas analysis system (IRGA). Custom-made collars that fit around each tree allow stem respiration to be measured (the IRGA is fitted to the collar, enabling the air inside it to be sampled). Other collars can be placed on the ground to demarcate specific areas of soil for sampling – again, the IRGA is used to sample the air inside the collar over a period of time.

CONTEMPORARY CASE STUDY: WYTHAM WOODS – THE CARBON CYCLE OF A TEMPERATE FOREST ECOSYSTEM IN THE UK

Wytham Woods (51°46'N 1°20' W) is a 400 ha temperate deciduous forest near Oxford, UK.

- The most common canopy trees are oak (*Quercus robur*), ash (*Fraximus excelsior*), sycamore (*Acer pseudoplatanus*) and beech (*Fagus sylvaticus*).

- The understorey contains shrubs such as hazel (*Corylus avellana*), hawthorn (*Crataegus monogyna*) and blackthorn (*Prunus spinosa*), and ground flora includes bluebell (*Endymion non-scripta*) and dog's mercury (*Mercurialis perennis*).

The woodland is relatively undisturbed, and has been the focus of ecological research based at Oxford University for many years (see Savill, *et al.* 2010). The climate is relatively mild, and can be classified as 'maritime', with a relatively narrow annual temperature range featuring warm (but not extremely hot) summers and cool (but not excessively cold) winters.

▲ **Figure 4.9** Wytham Woods, UK

Detailed studies have been carried out at Wytham Woods on the carbon cycle, at seasonal and annual scales. These investigations have given a detailed understanding of carbon cycling at the local scale, and have enabled comparisons to be made with similar studies in other forests with different climates or vegetational profiles. Why is it important to gain such an understanding? As well as allowing fluxes of carbon into and from the atmosphere to be quantified, and models of carbon movement to be formulated, such data allows scientists to better understand global climate change (Chapter 5, pages 140–3).

In order to understand the storages and flows of carbon in a forest ecosystem, the productivity and rate of respiration of each component of the forest must be measured to obtain biometric data. NPP can be measured in leaves, the woody parts of trees, below ground (e.g. roots), leaf litter, woody debris and reproductive organs (i.e. fruit and seeds). Respiration can be measured in leaves, stems, soil and leaf litter (due to activity of decomposers). The growth of trees was measured monthly using a subsample of 280 trees from a plot total of 466 over 10 cm DBH (diameter at breast height – a standard method of expressing the diameter of the trunk). Biometric and chamber measurements were made within the study area.

Eddy covariance measurements were taken from a tower within the sampling area, 25 m above the forest floor.

Chamber measurements and eddy covariance operate at different scales. Chamber measurements are made at very small scales whereas eddy covariance measurements are usually interpreted to represent average flux over relatively large areas.

Results and conclusions

Over the 24-month study period, eddy covariance data estimated that annual GPP = 21.1 MgC ha^{-1} yr^{-1}, R_e = 19.8 MgC ha^{-1} yr^{-1} and net ecosystem productivity (NEP) = 1.2 MgC ha^{-1} yr^{-1}.

Estimates from eddy covariance techniques were compared with the biometric measurements:

GPP = 26.5 ± 6.8 MgC ha^{-1} yr^{-1},
R_e = 24.8 ± 6.8 MgC ha^{-1} yr^{-1}

and

NEP = 1.7 MgC ha^{-1} yr^{-1}

Although there were some differences between the top-down and bottom-up approaches, eddy covariance estimates for both GPP and R_e are relatively closely matched within the biometric data (uncertainties within biometric measurements taken into account), suggesting that either technique can be used with confidence to estimate carbon budgets.

The Wytham Woods site was estimated to be a substantial carbon sink of 1.2 MgC ha^{-1} yr^{-1} between May 2007 and April 2009. Over the two years, GPP remained similar for equivalent months in different years whereas NEP varied:

difference in seasonal NEP was predominantly thought to be caused by changes in ecosystem respiration.

GPP and R_e values were generally higher than other broad-leaved temperate deciduous woodlands, possibly due to the influence of the UK's maritime climate, or the specific species composition of the site.

Figure 4.10 shows a summary of data compartmentalised and covering specific aspects of forest ecology, using results from both bottom-up and top-down approaches. This figure gives a clear indication of the complexity and magnitude of carbon fluxes at the local scale. Units for figure 4.10 are MgC ha^{-1} yr^{-1}.

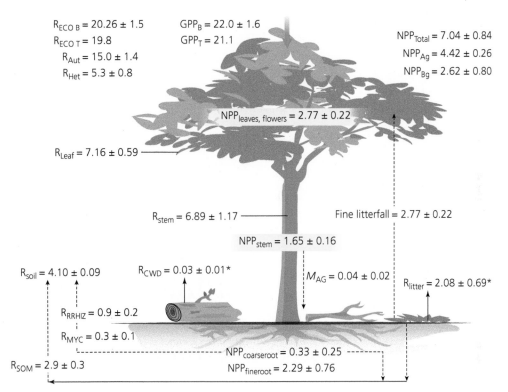

$R_{ECO\,B} = 20.26 \pm 1.5$
$R_{ECO\,T} = 19.8$
$R_{Aut} = 15.0 \pm 1.4$
$R_{Het} = 5.3 \pm 0.8$

$GPP_B = 22.0 \pm 1.6$
$GPP_T = 21.1$

$NPP_{Total} = 7.04 \pm 0.84$
$NPP_{Ag} = 4.42 \pm 0.26$
$NPP_{Bg} = 2.62 \pm 0.80$

$NPP_{leaves,\,flowers} = 2.77 \pm 0.22$

$R_{Leaf} = 7.16 \pm 0.59$

$R_{stem} = 6.89 \pm 1.17$

Fine litterfall $= 2.77 \pm 0.22$

$NPP_{stem} = 1.65 \pm 0.16$

$R_{soil} = 4.10 \pm 0.09$

$R_{CWD} = 0.03 \pm 0.01*$

$M_{AG} = 0.04 \pm 0.02$

$R_{litter} = 2.08 \pm 0.69*$

$R_{RRHIZ} = 0.9 \pm 0.2$

$R_{MYC} = 0.3 \pm 0.1$

$R_{SOM} = 2.9 \pm 0.3$

$NPP_{coarseroot} = 0.33 \pm 0.25$
$NPP_{fineroot} = 2.29 \pm 0.76$

▲ **Figure 4.10** The carbon cycle of Wytham Woods, constructed from the measured components. B indicates measurements taken from biometric (bottom-up) data and T indicates data taken from the tower (eddy-covariance/top-down); R indicates respiration, with $R_{ECO\,B}$ = respiration recorded from biometric data, $R_{ECO\,T}$ = respiration recorded from tower, R_{AUTO} = autotrophic respiration, R_{Het} = heterotrophic respiration; NPP_{Ag} indicates above-ground, NPP_{Bg} below-ground NPP, and NPP_{Total} the combined NPP from both above and below-ground; CWD = coarse woody debris; NPP_{LITTER} = canopy litter-fall (leaves, flowers, fruit) and F_{LITTER} = the fraction entering the soil (rather than respired in the litter layer); M_{AG} = the mean annual production of above-ground CWD over a three-year period; R_{RRHIZ}, R_{MYC}, and R_{SOM} are respiration from roots and the rhizosphere (a region of the soil that is affected by secretions from root and associated soil microorganisms), mycorrhizas (fungi that have a symbiotic relationship with tree roots) and soil organic matter decomposition, respectively, which sum to total soil respiration, R_{SOIL}

Chamber experiments are used to measure NPP, using light and dark measurements. Measurements taken in the light are direct measurement of NPP because they balance release of CO_2 to the chamber from respiration and uptake by photosynthesis (assuming there is vegetation in the chamber). Measurements taken in the dark (with a cover on the chamber) are respiration because photosynthesis is assumed to be zero.

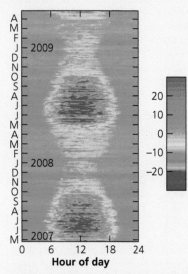

▲ **Figure 4.11** The net ecosystem exchange (NEE) 'fingerprint' of Wytham Woods showing diurnal (x-axis) and seasonal (y-axis) CO_2 fluxes

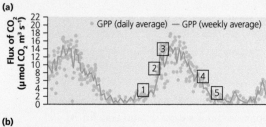

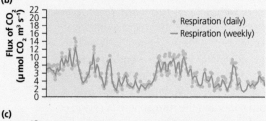

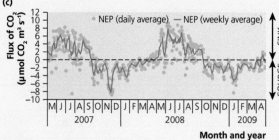

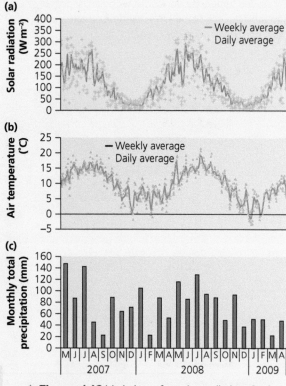

▲ **Figure 4.12** Variation of **a** solar radiation, **b** air temperature, **c** monthly total precipitation over the study period

◀ **Figure 4.13** Carbon fluxes from Wytham Woods showing **a** gross primary productivity (GPP), **b** ecosystem respiration (R_e), **c** net ecosystem productivity (NEP). Dots indicate daily mean values, lines trace weekly mean values. Specific stages of GPP are indicated (graph a, labels 1–5). Normal convention is that carbon sequestration is negative (a loss of carbon from the atmosphere) and a carbon loss from the biosphere is positive.

(a) Using data from Figure 4.11, describe changes in the seasonal and diurnal patterns of NEE over the study period.

GUIDANCE

This question asks you to describe (not explain) patterns in two sets of graphs. Describe overall trends.

(b) Using data from Figures 4.11 and 4.12, explain the observed pattern of carbon dioxide fluxes at Wytham Woods.

GUIDANCE

In your analysis, focus on interpreting (not describing) results and explaining observations you have made in Q1. Patterns in Figure 4.11 can be explained using information in Figure 4.12. Look for the big picture or overall trends, and use biological knowledge to explain trends and differences. You should focus on the degree of change and the relation between sets of data. Make sure you use specific language, for example 'increase' rather than 'change'. When describing graphical trends, do not just identify the basic pattern, but recognise more subtle points such as changes in rate. In Figure 4.11, dark blue areas represent maximum CO_2 sequestration, and red areas periods of net carbon dioxide efflux from the forest. Use your knowledge of photosynthesis, respiration and their interrelationship in determining NPP to interpret and explain the data. Think about seasonal and diurnal (i.e. daily) changes in solar radiation and how these may affect photosynthetic rates.

(c) Describe and explain the general trends seen in each graph (4.13a–c).

GUIDANCE

When describing patterns, refer to specific values in data (i.e. at minimum or maximum values) as well as overall trends. Look at different temporal patterns in different sets of data, e.g. respiration rates compared to photosynthetic activity. Once again, use scientific knowledge to explain observed patterns. What explanation is there for fluctuations in rates of respiration? You may need to refer to other graphs in this section to fully explain the data. What is the relationship between GPP, respiration and NEP? How are these relationships shown in the graphs?

(d) Explain the changes in GPP at the specific stages indicated on graph 4.13a (labels 1–5).

GUIDANCE

The question highlights key changes in rates of photosynthesis during a yearly cycle. A brief explanation is needed for each stage. Focus on the process that determines GPP, and how this varies throughout the year.

 KEY TERMS

Net ecosystem exchange (NEE) The net exchange of carbon between the atmosphere and vegetation, calculated by subtracting carbon losses in **heterotrophic** respiration from the amount of carbon produced by NPP. Negative NEE indicates net sequestration of carbon.

Heterotrophic A term relating to organisms that cannot photosynthesise and must obtain their food and energy by taking in organic substances, usually plant or animal matter. All animals, protoctistans, fungi and most bacteria are heterotrophic.

④ Evaluating the issue

▶ *Assessing the spatial and temporal variability of carbon system flows.*

Assessing the influence of climatic variations on global patterns of carbon cycling

The phrase 'climatic variations' can be interpreted in two different ways: spatial and temporal variations.

Firstly, there are spatial variations in climate to consider, at varying scales, in relation to ecosystem growth and carbon sequestration (in both vegetation and soils).

- Diagrams showing the global carbon cycle (Figures 4.1 and 4.7) give no indication of any spatial variability. In this chapter's discussion, we explore the important geographical theme of climatic *variability*:
  - ☐ Broad regional differences in climate affect carbon cycle flows in the Earth's different global biomes (Figure 4.14).
  - ☐ There are also more local-scale variations in climate to consider, for example in relation to relief and coastlines.

Secondly, climate varies temporally too:

- In most world regions, seasonal changes in climate bring marked differences in carbon cycling during different months of the year.
- Climate can also vary on longer timescales, with implications for carbon cycling. Year-to-year changes could be linked with operations of ENSO cycles, for instance (see page 16).
- There are also longer-term climate change processes and issues to consider.

It is also important to recognise that climate may be just one of a larger number of influences on carbon cycling in particular regional or local contexts. Other important factors include geology, relief, drainage and human activity.

Assessing the influence of *spatial* climatic variations on biome distribution and carbon cycling patterns

In the previous section we saw how inputs and outputs of carbon varied between different elements of a temperate forest, depending on rates of productivity and respiration in different parts of the system. NPP can be expected to vary greatly between different biomes, and therefore inputs and outputs of carbon, due to vegetation types varying according to climate and species composition. NPP is affected by factors that influence the rate of photosynthesis, namely:

- the amount of insolation (sunlight)
- the amount of rainfall (precipitation)
- the temperature.

The different biomes around the planet have very different climates. Figure 4.14 shows the distribution of biomes, with the variation and range of rainfall and insolation over a twelve-month cycle given for each.

Areas of highest productivity can expect to have the largest net absorption of carbon. Warmer temperatures speed up enzyme reactions that drive photosynthesis, although temperatures that are too hot can lead to denaturation of enzymes. Rainforests have the highest NPP because they are found between the Tropics of Cancer and Capricorn (23.5° N and S of Equator) where rainfall is high (over 2500 mm yr^{-1}), insolation is constant throughout the year and temperatures are warm. The high NPP means that rainforest has a complex structure with a number of layers from ground level to canopy, with emergent trees up to 50 metres and lower layers of shrubs and vines.

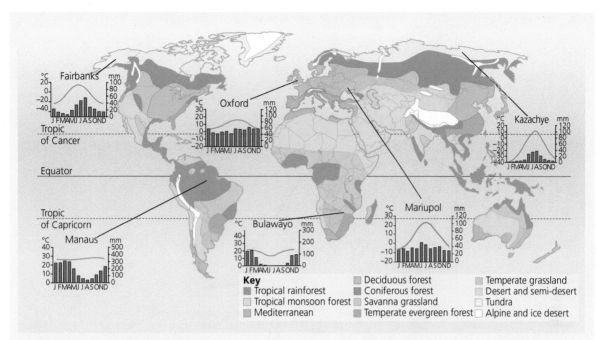

▲ Figure 4.14 The global biome map

Temperate ecosystems vary from deciduous and evergreen forests to grasslands. Although temperate forests are highly productive for part of the year, variation in insolation due to seasonal variation limits their overall NPP and thus carbon sequestration. Rainfall and temperature are also seasonal, which further reduces overall NPP compared to tropical rainforests. Lower levels of NPP mean less stored chemical energy at the producer level. Rainfall is sufficient in temperate forest areas to establish forest (500–1500 mm yr^{-1}) rather than grassland.

In colder ecosystems found in the northern hemisphere, such as tundra, water is locked up as ice and is not available to plants, reducing productivity and carbon fluxes. In highly stressful environments, with very low temperature or rainfall, only mosses and lichens may be able to survive. Measurements taken in arctic tundra at Barrow, Alaska, suggest that overall the ecosystem is a net sink for carbon, although lowering the water table from the soil surface to −5 cm had significant effects and decreased net ecosystem carbon storage.

The type of stable ecosystem that will emerge in an area is predictable based on climate. Information about the relative contributions of two climatic factors, precipitation and temperature, can be used to predict the type of stable ecosystem that can be expected in a given area. A climograph is a graphical model that shows the relationship between temperature, precipitation and ecosystem type and shows the likely stable ecosystems that are found under specific climatic conditions (Figure 4.15).

Carbon can be kept in stores within ecosystems (in biomass, soil or leaf litter), moving between different storages as transfers. Biological material is made from carbon-based molecules (Chapter 4, page 100) and so as nutrients flow between stores, so too does carbon. For example, leaf-fall from trees decomposes and returns carbon to the soil. The movement of nutrients (and therefore carbon) can be shown in diagrammatic form (see Figure 4.16). The size of the nutrient stores (biomass, litter and soil) is proportional to the quantity of the nutrients stored, and the thickness of the transfer arrows is proportional to the amount of nutrients transferred.

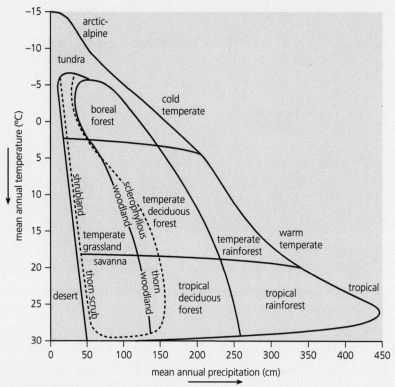

▲ **Figure 4.15** Whittaker's climograph

System flows can be shown for different biomes, indicating spatial variability in the intensity of inputs and outputs from ecosystem stores (Figure 4.17). These nutrient cycles show us the impact of climate variations on rates and volumes of carbon cycling.

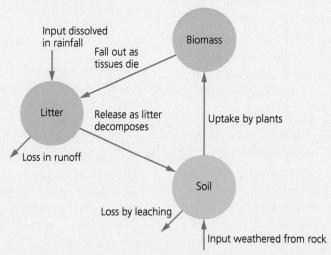

▲ **Figure 4.16** A model of the nutrient cycles in ecosystems; circles represent nutrient stores and arrows nutrient transfers

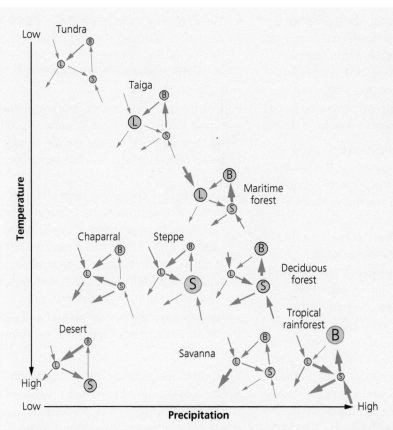

▲ **Figure 4.17** Nutrient cycling models for ecosystems across the world (after Gersmehl, 1976). Note: this figure, and also Figure 4.15, use temperature and precipitation to indicate levels of GPP, rather than carbon sequestration *per se*. The largest terrestrial carbon store is found in peatlands, which tend to fall at the top of the diagram (with low temperature and moderate to high rainfall). Peatlands have very high soil storage due to low respiration rates in waterlogged and anaerobic soils.

Assessing the influence of *temporal* climatic variations on carbon cycling

If carbon fluxes vary within an ecosystem, and depend on diurnal and seasonal changes of climate, temperature, sunlight and foliage (see Wytham Woods case study, page 114), then to what extent do system flows vary seasonally at the global scale? Biomes with consistent levels of solar radiation throughout the year (i.e. those found on the Equator) can be expected to have more constant rates of GPP and NPP than

seasonal biomes, such as temperate forest. In the Wytham Woods case study, we saw how rates of GPP, respiration and NEP vary throughout the year, and are affected by seasonal changes in precipitation, temperature and solar radiation. Variation in water supply throughout the year (Chapter 2) can be expected to affect rates of photosynthesis. Some biomes show clear seasonal changes, for example temperate deciduous woodland and savannas, whereas tropical rainforests and coral reefs show little seasonal variation (Chapter 1). In grasslands, warm, humid conditions in autumn ensure rapid decomposition of dead grass matter and the quick

Pearson Edexcel
AQA
OCR
WJEC/Eduqas

release of CO_2, and as a result, the litter store is small. Exchanges of carbon between atmosphere, biosphere and soil vary greatly according to season. In winter, the grasses die back to their roots and photosynthesis ceases. Plants can continue to respire through their roots, however, especially once the soil begins to warm in spring.

Clearly, temperate world regions experience seasonal changes in carbon cycling. But to what extent may tropical biomes experience similar changes? Research has shown that influxes and effluxes of carbon in equatorial biomes can be affected by changes in climatic conditions. For example, in the Amazon basin, drought conditions change the ecosystem from a net sink of carbon to a net source. During a dry year (2010) the Amazon basin lost 0.48 ± 0.18 petagrams of carbon per year ($PgCyr^{-1}$) but was carbon neutral ($0.06 \pm 0.1 PgCyr^{-1}$) during a wet year (2011). If carbon losses from fire were taken into account, research suggests that vegetation was a net carbon sink of $0.25 \pm 0.14 PgCyr^{-1}$: this result is consistent with the mean long-term intact-forest biomass sink of $0.39 \pm 0.10 PgCyr^{-1}$ estimated from previous forest studies. Observations from Amazonian forest plots suggest the suppression of photosynthesis during drought was the main cause for the 2010 sink neutralisation and that moisture has an important role in determining the Amazonian carbon balance.

In Chapter 3, we explored how different regions experience large variability in water supply – this variability will affect rates of photosynthesis and productivity. For example, India gets 90 per cent of its rainfall during the summer monsoon season – at other times rainfall over much of the country is very low (Chapter 3, page 77).

Year-to-year or longer-term temporal variations in climate and carbon cycling

Some drought is seasonal, whereas many parts of the world experience naturally occurring climatic variability lasting for years or even decades:

- The El Niño southern oscillation (ENSO) is a naturally occurring climatic phenomenon in the Pacific Basin (Chapter 3) which occurs every three to seven years and usually lasts for eighteen months.
- During an El Niño event, cool water normally found along the coast of Peru is replaced by warmer water. At the same time, warmer water near Australia and Indonesia becomes cooler.
- As a result of these and other changes, very dry conditions arrive in parts of south-east Asia, India, eastern Australia, south-eastern USA and Central America. In these contexts, precipitation transfers from the atmosphere to the land are greatly reduced for a period of time. In contrast, rainfall increases across the east-central and eastern Pacific Ocean, including the Californian coastline.

Changes in temperature and precipitation affect ecosystem productivity (pages 116–9) and so these longer-term variations in climatic conditions affect carbon fluxes and stores over long time periods. Changes in rainfall and aridity due to climate change (Chapter 5) also have long-term effects on the carbon cycle. Warmer air can hold more moisture, so global warming is expected to result in large changes in precipitation. Broadly, areas that already have a lot of rain can be expected to experience more, while areas that are already arid may suffer lower rainfall. Rainfall patterns are also likely to change, with an increase in very heavy, episodic downpours, perhaps punctuated by longer periods of drought. Areas viewed as being at high risk of increasing aridity include desert fringes in Australia, China, the USA and the Sahel region (a long strip of semi-arid drylands including parts of Sudan, Chad, Burkina Faso and Niger). Climate data for the Sahel suggests that a long-term reduction in rainfall may be taking place.

Arriving at an evidenced conclusion

There is clear spatial and temporal variability of carbon system flows at a global scale. Differences in carbon stores in different biomes (Figure 4.18 provides another view of this) reflect different climatic conditions which determine the magnitude of carbon fluxes into ecosystems compared to movement from the ecosystem into the atmosphere, which in turn determines GPP and NPP.

As well as variation between biomes at the global scale, there are differences between flows within each ecosystem, determined by the duration and intensity of precipitation, rate of decomposition, degree of weathering and temperature. Variation also occurs seasonally and at longer timescales. In broad terms, variation in system flows of carbon are determined by global variation in precipitation, insolation and temperature. By measuring climatic variables, an understanding of spatial variability of carbon flows can be garnered – details missing from generalised models of the carbon cycle.

World biomes and carbon storage

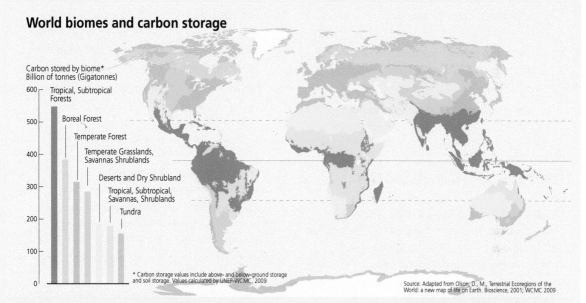

▲ **Figure 4.18** Carbon stores in different biomes. The world's forests store about 650 billion tons of carbon; more carbon than is found in the atmosphere. When forests are burnt or destroyed, the carbon is released as CO_2. When forests grow, either through expansion of forest area or because old forests become denser and more carbon-rich, they convert carbon dioxide to carbon in the form of wood and other biomass via the process of photosynthesis. Figures show vegetation carbon storage. NB: Soil carbon is about three times that in forests and half of that is in northern hemisphere peatlands: the map would look very different if it accounted for the whole terrestrial carbon store not just live plants.

Chapter summary

✔ The global carbon cycle maintains a balance between the terrestrial and oceanic stores of carbon. The distribution and size of stores varies between different biomes, as do the flows between them.

✔ Carbon flows and processes include those between:

☐ land and atmosphere at the local, short-term scale, including fossil fuel combustion, carbon sequestration and the processes of photosynthesis, respiration and decomposition

☐ ocean and atmosphere through the processes of absorption by biota, diffusion into and out of oceans

☐ land and oceans at the continental scale through the processes of weathering, river transport, indirect movement via the water cycle and carbon sequestration in sediments over millions of years.

✔ The carbon cycle can be studied at the local scale using eddy covariance and biometric data. Such studies indicate the fluxes between different elements of an ecosystem and can be used in predicting the potential effects of climate change.

✔ The size of flows of carbon between various stores in different biomes, for example tropical rainforest and temperate forest, are determined by temperature, precipitation and light.

✔ Marked seasonal changes can occur in carbon cycling, along with variations occurring over longer timescales.

Refresher questions

1 Define the following processes of the carbon cycle: photosynthesis; respiration; decomposition; fossilisation.

2 Draw an annotated diagram of the global carbon cycle showing its main stores and flows.

3 Using examples, explain the difference between fast carbon cycle flows between the land and the atmosphere and slow carbon cycle flows between the land and the ocean.

4 Explain the concept of mass balance in relation to the carbon cycle.

5 Outline the differences between net primary productivity and gross primary productivity.

6 Explain how NPP, GPP and carbon fluxes can be measured for different ecosystems using eddy covariance and biometric measurements.

7 Explain the main features of the global distribution of biomes.

8 Describe how and explain why annual carbon flows vary for **three** different biomes.

9 Suggest reasons why patterns of carbon flow and storage may vary within **one** named biome.

Discussion activities

1 Working in pairs, produce a carbon cycle diagram for an ecosystem near to where you live. Annotate flows to indicate processes involved.

2 Working in small groups, discuss how carbon cycles may vary spatially and temporally at a local scale, using a case study of your choice. Produce a poster of your findings and invite the rest of your class to comment and ask questions.

3 In small groups, discuss your personal experiences of visiting places (locally and globally) where different ecosystems are found. Can you suggest how the carbon storage and flow characteristics of these environments might differ from other places and environments?

4 In pairs, carry out online research about an aspect of carbon cycling you are keen to find out more about. Once each pair of students has completed their research, they can discuss the findings with the rest of the class.

5 Discuss the view that, when an area is being changed or redeveloped for commercial purposes, it is equally desirable to minimise water cycle and carbon cycle changes.

FIELDWORK FOCUS

Carbon cycle dynamics offers plenty of opportunities for ecological investigations.

A *Comparing soil organic content from different areas of an ecosystem.* Areas can be selected according to their canopy cover (i.e. open areas vs. more closed areas), vegetation type (tree vs. shrub) or location across an ecological gradient (i.e. deep forest through to forest edge), for example. The organic (i.e. carbon-containing) content of soil can be measured by first drying the soil and then burning off the organic content in an oven.

B *Analysing the organic content of stream water.* Samples of water can be taken from different positions in a stream for analysis in the lab. Comparing areas with more overhanging vegetation to more open areas may provide a good comparison. Water samples are poured through filter paper to remove solid organic sediments. The sediment is then placed in a crucible and burnt using a Bunsen burner in a well ventilated room. The difference between mass before burning and mass after burning is the organic content. The water from each sample can be further tested using a colorimeter: dissolved organic content contained within the water will alter the colour of the water. The absorbance of samples from different parts of the river can be compared to determine differences in dissolved organic content.

C *Estimating tree carbon stores in contrasting ecosystems.* The carbon stored in tree trunks can be estimated for different forest types. One possibility is to estimate the circumference and height of trees, and from this calculate tree biomass and carbon content. The following sites have suggestions as to how to do this:

- www.forestry.gov.uk/pdf/Revised-biomass-equations-27Jan2014.pdf/$FILE/Revised-biomass-equations-27Jan2014.pdf
- www.geography-fieldwork.org/a-level/water-carbon/carbon-cycle/method/

Further reading

Fenn, K., Malhi, Y., Morecroft, M., Lloyd, C. and Thomas, M. (2015) 'The carbon cycle of a maritime ancient temperate broadleaves woodland at seasonal and annual scale', *Ecosystems* 18, Issue 1, pages 1–15, doi:10.1007/s10021-014-9793-1

Gatti, L. V. *et al.* (2014) 'Drought sensitivity of Amazonian carbon balance revealed by atmospheric measurements', *Nature* 506, pages 76–80. doi:10.1038/nature12957

Malhi, Y., Meir, P. and Brown, S. (2002) 'Forests, carbon and global climate', *Philosophical Transactions of the Royal Society of London* 360, pages 1567–91

Savill, P., Perrins, C., Kirby, K. and Fisher, N. (eds), (2010) *Wytham Woods: Oxford's Ecological Laboratory*, Oxford University Press

Thomas, M. V. *et al.* (2011) 'Carbon dioxide fluxes over an ancient broad-leaved deciduous woodland in southern England', *Biogeosciences* 8, pages 1595–1613, doi:10.5194/bg-8-1595-2011

CHAPTER 5

Fossil fuel use and the implications of climate change

A balanced carbon cycle is important in sustaining Earth systems but is increasingly altered by human activities. Changes to the most important stores of carbon and carbon fluxes are a result of physical and human processes. Reliance on fossil fuels has caused significant changes to carbon stores and flows, and has contributed to climate change resulting from anthropogenic carbon emissions. This chapter:

- investigates the issue of energy security alongside fossil fuel use trends and patterns
- explores anthropogenic and natural climate change processes
- analyses the implications of climate change for physical systems and life on Earth
- evaluates the 'Anthropocene' concept in relation to carbon cycle dynamics.

KEY CONCEPTS

Anthropocene The idea that there are no truly natural environments, and that we are entering a new geological epoch where human activities dominate earth surface processes. Human activities have modified physical systems by their carbon emissions, for example, even in areas where there are relatively few people, such as high latitudes and the oceans. In other environments that are conducive to human habitation, such as river valleys, the impact of human activity has been considerable (see also page 20).

Energy security The uninterrupted availability of energy sources at an affordable price.

Negative feedback The process by which a system acts to decrease the effects of the original change to the system (self-regulation).

Positive feedback A change in a system which leads to an increasing deviation from the original (also known as vicious circle, snowball effect).

 Energy security and fossil fuel use trends and patterns

▶ *What is the relationship between fossil fuel use in different countries and their energy security status?*

 KEY TERM

Energy security The uninterrupted availability of energy sources at an affordable price.

Energy security is a key goal for countries, with most relying on fossil fuels. Energy security depends on an adequate, reliable and affordable supply of energy that provides a degree of independence. An inequitable availability and uneven distributions of energy sources may lead to conflict. Energy security refers to a country's ability to secure all its energy needs, whereas energy insecurity refers to a lack of security over energy sources.

According to the analyst Chris Ruppel (2006), the period from 1985 to 2003 was an era of energy security, and since 2004 there has been an era of energy insecurity. He claims that following the energy crisis of 1973 and the First Gulf War (1990–1), there was a period of low oil prices and energy security. However, insecurity has since arisen for a number of reasons, including:

- increased demand, especially by emerging economies including China, India, Brazil and Indonesia, putting pressure on existing supply lines.
- decreased reserves as supplies are used up, meaning that countries need to look for alternative supplies.
- geopolitical development: countries such as Venezuela, Iran and Russia have 'flexed their economic muscle' in response to their oil resources and the decreasing resources in the Middle East and North Sea, leading to scarcity and unpredictability of oil resources.
- global warming and natural disasters such as Hurricane Katrina, which have increased awareness about the misuse of energy resources, leading to movement away from fossil fuel resources.
- terrorist activity (e.g. in Nigeria and Iraq), resulting in restricted access to reserves.
- the conflict between Russia and Ukraine. Europe is dependent on Russia for gas, and political conflict over Ukraine may result in restricted gas supplies.

Energy insecurity can cause and be the result of geopolitical tension. For most consumers, a diversified energy mix is the best policy, rather than depending on a single supplier.

 KEY TERM

Global warming An increase in average temperature of the Earth's atmosphere.

Factors affecting the choice of energy generation

There are many important factors to consider in the choice of energy resources by societies. These include the following:

- The availability and reliability of supply – the UK used to rely on coal, then it became oil-dependent, but it has limited potential for solar or geothermal energy.
- Sustainability of supply – there are perhaps 40 years' worth of oil and 140 years' worth of coal remaining according to some estimates, but an infinite supply of solar and geothermal energy exists in the world.
- Scientific and technological development – developing countries use less energy and more basic energy (e.g. fuelwood) whereas developed countries and, increasingly, emerging economies, use more energy and more expensive forms (e.g. nuclear and oil).
- Political factors – in 1973, the Organisation of Petroleum Exporting Countries (OPEC) raised the price of oil four-fold, causing other countries to develop their own cheaper resources.
- Economic factors such as the relatively high cost of production and distribution mean that nuclear power or tidal energy may be too expensive for many countries.

- Cultural attitudes – increased awareness of the problems of global warming or the risks associated with nuclear power may cause nations to change their energy choices.
- Environmental factors – certain climates allow for the use of certain types of energy such as solar or wind power; colder climates require more heating, warmer climates more air-conditioning.

The choice of energy sources adopted by different countries often has an historical basis. Large oil, coal (Figure 5.1) and gas reserves (Figure 5.2) in certain countries (e.g. the UK) made fossil fuels an obvious choice for exploitation in those countries. Energy generation may also depend on economic, cultural, environmental and technological factors.

▶ **Figure 5.1** Coal extraction in the UK. Coal mining techniques developed in the eighteenth century and production increased substantially in the nineteenth century, helping to fuel the industrial revolution. By the early 1980s, many pits were considered uneconomic compared to North Sea oil and gas extraction, and action by mining unions caused conflict with the government of the day which accelerated mine closures. The last deep coal mine operation in the UK, Kellingley Colliery, closed on 18 December 2015

▶ **Figure 5.2** Extraction of oil and gas from the North Sea took off as a commercial operation in the 1960s, but it was the oil crisis of 1973, causing world oil prices to quadruple, that led to such offshore production becoming more economically viable

Oil use in developed countries is almost 50 per cent greater than in developing countries, and fossil fuels in developed countries account for 85 per cent of energy use as opposed to 58 per cent in developing countries. There are various explanations for these observed patterns. Oil is used extensively to produce petroleum products, and difference in oil use between developing countries and developed countries can be explained by the more prevalent use of cars in developed countries. Biomass is very important in developing countries as fuel for cooking, whereas developed countries use gas or electricity (i.e. fossil fuels).

Factors affecting the demand for and supply of fossil fuels

As might be expected, global variations in fossil fuel supply occur for a number of reasons. These can be broadly subdivided into physical, economic and political factors. Table 5.1 shows examples for each of these groupings. The combination of factors operating in each country can vary significantly.

Physical	Economic	Political
• Deposits of fossil fuels are only found in a limited number of locations. • Large power stations require flat land and geologically stable foundations.	• The most accessible, and lowest-cost, deposits of fossil fuels are invariably developed first. • Onshore deposits of oil and gas are usually cheaper to develop than offshore deposits. • In poor countries, Foreign Direct Investment (FDI) is often essential for the development of energy resources. • When energy prices rise significantly, companies increase spending on exploration and development.	• International agreements such as the Kyoto Protocol can have a considerable influence on the energy decisions of individual countries. • Governments may insist on energy companies producing a certain proportion of their energy from renewable sources. • Legislation regarding emissions from power stations will favour the use of, for example, low-sulphur coal, as opposed to coal with a high sulphur content.

▲ **Table 5.1** Global variations in fossil fuel use due to physical, economic and political factors

The key factor in supply is energy resource endowment. Some countries are relatively rich in domestic energy resources, while others are lacking in such resources and heavily reliant on imports. However, resources by themselves do not constitute supply. Capital and technology are required to exploit resources.

The use of energy in all countries – and thus carbon-emitting fossil fuels – has changed over time due to a number of factors:

- Technological development – for example: (a) nuclear electricity has only been available since 1954, (b) oil and gas can now be extracted from much deeper waters than in the past, and (c) renewable energy technology is advancing steadily.

- Increasing national wealth – as average incomes increase, living standards improve, which involves the increasing use of energy and the use of a greater variety of energy sources.
- Changes in demand – at one time, all of Britain's trains were powered by coal and most people also used coal for heating in their homes. Before natural gas was discovered in the North Sea, Britain's gas was produced from coal (coal gas).
- Changes in price – the relative prices of the different types of energy can influence demand. Electricity production in the UK has been switching from coal to gas over the last twenty years, mainly because power stations are cheaper to run on natural gas.
- Environmental factors/public opinion – public opinion can influence decisions made by governments. People today are much better informed about the environmental impact of fossil fuel energy sources than they were in the past.

Oil: global patterns and trends

Oil is the most important of the non-renewable sources of energy. It is a compact and portable source of energy that is relatively easy to transport (i.e. it has a well-established global transport structure) and store. 'Carbon footprint' is a term that refers to the amount of carbon dioxide released into the atmosphere resulting from the activities of a particular individual, organisation or community, and so is an indicator of its contribution to global warming (page 140).

A standard unit for measuring carbon footprints is CO_2e, or 'carbon dioxide equivalent'. Burning fossil fuels releases mainly carbon dioxide, but other greenhouse gases (GHG) are also emitted, e.g. methane (CH_4) and nitrous oxide (N_2O). This unit allows a single number to be used rather than a carbon footprint consisting of lots of different greenhouse gases. It measures the impact of each different greenhouse gas in terms of the amount of CO_2 that would create the same amount of warming. Gigawatt hour (GWh) is a unit of electrical energy. To measure the amount of GHG emitted per hour of electricity generation, the unit CO_2e/GWh is therefore used.

Oil produces less carbon dioxide than coal (733 compared to 888 tonnes CO_2e/GWh) and so has a lower carbon footprint (although not as low as gas – see page 134). Even though investment in new sources of energy is increasing rapidly, the global economy still relies on oil to a considerable extent.

However, its disadvantages have gained increasing recognition in recent decades, such as its carbon footprint compared to other sources of energy generation. Serious oil spills have occurred from super tankers and

◀ **Figure 5.3** On 22 April 2010, a semi-submersible drilling rig, the Deepwater Horizon, sank in the Gulf of Mexico after an explosion on the vessel. When the vessel sank it detached a 5000-foot pipe that connected the oil-wellhead to the rig, and began leaking oil. A leak also developed in the wellhead itself. An estimated 5000 barrels (200,000 gallons) of oil per day began leaking into the Gulf. The disaster coincided with the period when many threatened species were moving to the Gulf to spawn, migrate and feed, including Atlantic blue fin tuna, Kemp's ridley sea turtles, loggerhead sea turtles, piping plovers and sperm whales

pipelines, leading to environmentally damaging pollution (Figure 5.3). Other environmental damage comes from the use of strip-mines, in the form of tar sands (Figure 5.4). Political instability of some major oil-producing countries raises concern about the vulnerability of energy pathways. There are also concerns over the variability in the price of oil, and that 'peak oil' may not be far away (see below).

◀ **Figure 5.4** Tar sands oil extraction. Tar sands are deposits of heavy black viscous oil (bitumen) combined with clay, sand and water. The tar cannot be pumped from the ground and so must be mined, usually using open-pit techniques where the surface above the deposits is removed to expose the tar sands beneath. Much of the world's oil is in the form of tar sands. The largest deposits are in Canada (Alberta) and Venezuela, with much of the rest found in the Middle East. In the USA most tar sands are found in Eastern Utah. Both the mining and processing of tar causes a variety of environmental impacts, such as greenhouse gas emissions, habitat disturbance, impacts on air quality and social and economic impacts on the local community

Pearson Edexcel

AQA

OCR

WJEC/Eduqas

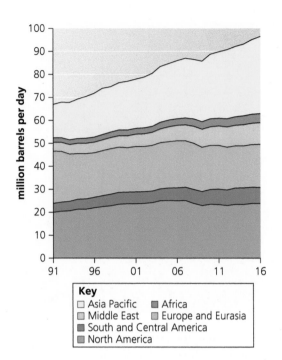

▶ **Figure 5.5** Oil consumption by world region, 1991–2016. The largest increase has been in the Asia Pacific region, which now accounts for 33.6 per cent of consumption. This region currently uses more oil than North America, which accounts for 24.6 per cent of the world total. In contrast, Africa consumed only 4 per cent of global oil. BP Statistical Review of World Energy 2016

Figure 5.5 shows the change in daily oil consumption by world region from 1991 to 2016. From just under 70 million barrels daily in 1991, global demand rose steeply to 97.8 million barrels a day in 2017. Satisfying such a rapid rate of increase in demand requires a high level of investment and exploration, and has environmental and other consequences.

World oil production grew by only 0.4 million barrels a day (b/d) in 2016: this was the slowest growth since 2013. Iran, Iraq and Saudi Arabia drove production in the Middle East, with production rising by 1.7 million b/d. Increased production in the Middle East was largely offset by declines in North America, Africa, Asia Pacific and South and Central America. Global oil consumption growth averaged 1.6 million b/d, above the ten-year average of one million b/d for the second successive year as a result of stronger than usual growth in the Organisation for Economic Co-operation and Development (OECD), with China and India providing the largest contributions to growth.

Region	Reserves/production ratio (years)
North America (USA, Mexico and Canada)	32.3
South and Central America	119.9
Europe and Eurasia	24.9
Middle East	69.9
Africa	44.3
Asia Pacific	16.5
World	50.6

▲ **Table 5.2** Oil reserves-to-production ratio at the end of 2016

Source: BP Statistical Review of World Energy, 2017

The price of oil increased sharply in the early years of the new century, causing major financial problems in many importing countries. It rose from US$10 a barrel in 1998 to more than US$130 a barrel in 2008, before falling back sharply in the global recession of 2008–9. As the global economy has slowly recovered, the price of oil has fluctuated considerably and was at a price of about US$40 a barrel in late 2015. In January 2018 the price of oil reached US$70 a barrel for the first time since December 2014. Crude oil prices will average US$62 a barrel in 2018 and are expected to average US$75 per barrel in 2019.

When will global peak oil production occur?

There has been growing concern about when global oil production will peak and how fast it will decline thereafter. Peak oil production refers to the year in which the world or an individual oil-producing country reaches its highest level of production, with production declining thereafter (Figure 5.6). Peak oil varies country by country. The peak of oil discovery occurred in the 1960s, and by the 1980s the world was using more oil than was being discovered. Since then the gap between use and discovery has been increasing, and many countries have passed their peak oil production. We depend on oil for many things: we use it for fuel, transport and heating, as a raw material in the plastics industry, and for fertiliser in food production. As oil production decreases after peak oil, so will all of these, unless we can find new materials and alternatives.

KEY TERM

Peak oil The point at which global oil achieves its maximum rate of production, after which production starts to decline.

There are concerns that there are not enough large-scale projects underway to offset declining production in well-established oil-production areas. The rate of major new oil-field discoveries has fallen sharply in recent years. It takes six years on average from first discovery for a very large-scale project to start producing oil. In 2010, the International Energy Agency expected peak oil production somewhere between 2013 and 2037, with a fall by 3 per cent a year after the peak. The United States Geological Survey predicted that the peak was 50 years or more away. However, in complete contrast, the Association for the Study of Peak Oil and Gas (ASPO) predicted in 2008 that the peak of global oil production could come as early as 2011, stating 'Fifty years ago the world was consuming 4 billion barrels of oil per year and the average discovery was around 30 billion. Today we consume 30 billion barrels per year and the discovery rate is now approaching 4 billion barrels of crude oil per year.'

ASPO's dire warnings have not (yet) materialised. This is at least partly down to new developments, particularly the rapid growth in production of shale oil and gas in the USA, which have changed the global energy situation. The current period of slow growth in the global economy has also eased the pressure on energy resources.

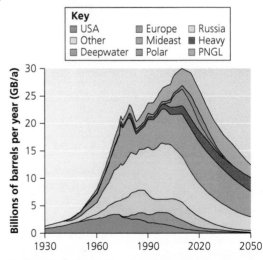

▲ **Figure 5.6** Actual and projected peak oil production for different countries and world regions

The geopolitical impact of changes in patterns and trends in oil

Energy security depends on resource availability – domestic and foreign – and security of supply. It can be affected by geopolitics, and is a key issue for many economies. Because there is little excess capacity to ease pressure on energy resources, energy insecurity is rising, particularly for non-renewable fossil fuel resources.

The Middle East is the major global focal point of oil exports. The long-running tensions that exist in the Middle East have at times caused serious concerns about the vulnerability of oil fields, pipelines and oil-tanker routes. The destruction of oil wells and pipelines during the Second Gulf War or the Iraq War (2003–11) showed all too clearly how energy supplies can be disrupted. Middle East oil exports are vital for the functioning of the global economy. Most Middle East oil exports go by tanker through the Strait of Hormuz, a relatively narrow body of water between the Persian Gulf and the Gulf of Oman. The strait at its narrowest is 55 kilometres wide. Roughly 30 per cent of the world's oil supply passes through the Strait, making it one of the world's strategically important chokepoints. Iran has at times indicated that it could block this vital shipping route in the event of serious political tension. This could cause huge supply problems for many importing countries. Concerns about other key energy pathways have also arisen from time to time.

Natural gas

Gas produces 499 tonnes CO_2e/GWh and so has a much lower carbon footprint than both oil and coal. Global production of natural gas increased from 2876.7 billion m^3 in 2006 to 3551.6 billion m^3 in 2016 (Table 5.3).

Region	2006	2016	% change	% share of market in 2016
North America	753.0	948.4	26	26.7
South and Central America	154.1	177.0	15	5.0
Europe and Eurasia	1042.2	1000.1	−4	28.2
Middle East	343.6	637.8	86	18.0
Africa	192.6	208.3	8	5.9
Asia Pacific	391.3	579.9	48	16.3
Total world	2876.7	3551.6	24	100.0

▲ **Table 5.3** Natural gas production by world region, 2006–16. Five world regions showed an increase in production. However, one of the largest producing world regions, Europe/Eurasia, showed a reduction in production between 2006 and 2016. The highest relative change was in the Middle East. The big increase in North American gas production was due to an increase in fracking. The increases in the middle east were in part an attempt to destabilize the North American fracking industry by driving down prices.

Source: BP Statistical Review of World Energy, 2017

KEY TERM

Fracking Hydraulic fracturing (fracking) is a process that releases gas from shale formations by pumping water, sand and chemicals into the rock.

On an individual country basis, natural gas production is dominated by the USA (21.5 per cent of the global total) and the Russian Federation (16.2 per cent). There is a very substantial gap between these two natural gas giants and the next largest producers, which are Iran (5.7 per cent), Qatar (5.1 per cent) and Canada (4.3 per cent). Global consumption of natural gas in 2016 was led by Europe and Eurasia (28.9 per cent), North America (27.7 per cent) and Asia Pacific (20.3 per cent).

During the period 2006–16, proven reserves of natural gas increased substantially. The global share of proven reserves in the Middle East and Southern and Central America fell slightly, while the share held in Europe and Eurasia increased. On an individual country basis, the largest reserves in 2016 were in Iran (18 per cent), the Russian Federation (17. 3 per cent) and Qatar (13.0 per cent). In 2016, the global reserves-to-production ratio stood at 52.5 years.

As with oil, one of the advantages of natural gas is the well-established global network for trade (Figure 5.7). As you can see, trade in energy is an important component of global economic systems.

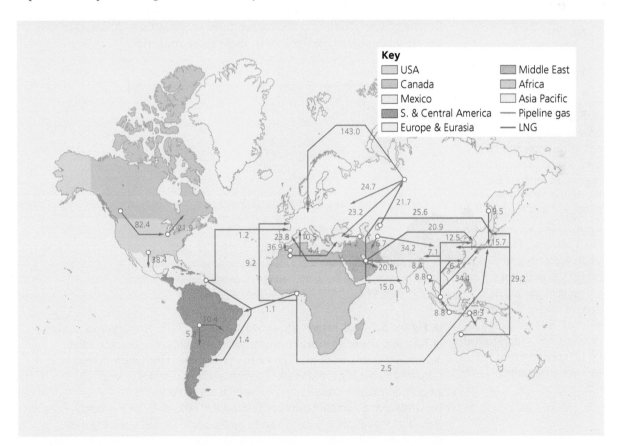

▲ **Figure 5.7** Trade movement of gas in 2016. Trade flows are represented in billion cubic metres. LNG = liquefied natural gas

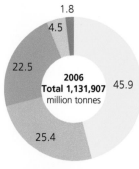

1.8
4.5
22.5
2006
Total 1,131,907
million tonnes
45.9
25.4

1.3 — 1.2
22.8
2016
Total 1,139,331
million tonnes
46.5
28.3

Key
- Asia Pacific
- Europe and Eurasia
- North America
- Middle East and Africa
- South and Central America

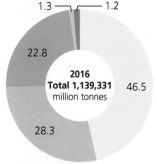

▲ **Figure 5.8** Distribution of proven coal reserves, 2006 and 2016, showing a fairly even spread between three regions: Europe/Eurasia, Asia Pacific and North America. By region, Asia Pacific holds the most proved reserves (46.5 per cent of total), with China accounting for 21.4 per cent of the global total. The US remains the largest reserve holder (22.1 per cent of total). Total global reserves, however, declined by 12.5 per cent over this ten-year time period

Coal

Coal produces an average of 888 tonnes CO_2e/GWh and so has the highest carbon footprint of any fossil fuel. Coal is responsible for 43 per cent of carbon dioxide emissions from fuel combustion, 36 per cent is produced by oil and 20 per cent from natural gas.

Coal production is dominated by the Asia Pacific region, accounting for 71.6 per cent of the global total in 2016. Much of this coal is produced in China, which alone mines 46.1 per cent of the world total. The next largest producing countries were the USA (10.0 per cent), Australia (8.2 per cent), India (7.9 per cent), Indonesia (7.0 per cent), and the Russian Federation (5.3 per cent). Consumption is led by Asia Pacific (73.8 per cent), Europe and Eurasia (12.1 per cent) and North America (10.4 per cent). China alone consumed 50.6 per cent of world coal in 2016.

Figure 5.8 shows the proven reserves of coal in 2006 and 2016. The global reserves-to-production ratio was 153 years in 2016 (Figure 5.9). Coal reserves, however, can become exhausted within a relatively short time period. In the nineteenth and early twentieth centuries, countries such as Germany, the UK and France were significant producers. Today, there are very few operational coal mines in these three countries.

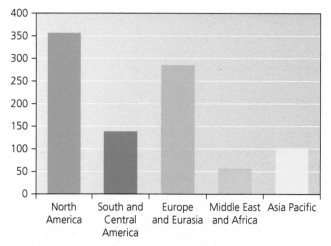

▲ **Figure 5.9** Coal: reserves-to-production ratios, 2016. A reserves-to-production, or R/P, ratio is used to measure the remaining amounts of a non-renewable resource, such as coal or gas. The ratio is expressed in time (years). The ratio is calculated by dividing the amount of known available reserves in an area by the amount of resource produced in one year (at the current rate), i.e. R/P = amount of known resource / amount used per year. An R/P ratio of 350, for example, does not mean that the resource will continue to be produced for 350 years and then suddenly run out: a resource will increase in production, peak, and then plateau before entering a declining phase

Extending the 'life' of fossil fuels

Coal gasification is the process of producing a mixture of carbon monoxide, hydrogen, carbon dioxide, methane, and water vapour from coal. At present, electricity from coal gasification is more expensive than that from traditional power plants, but if more stringent pollution laws are passed in the future this situation could change significantly.

Clean coal technology has developed forms of coal that burn with greater efficiency and capture coal's pollutants before they are emitted into the atmosphere. The latest 'supercritical' coal-fired power stations, operating at higher pressures and temperatures than their predecessors, can operate at efficiency levels 20 per cent above those of coal-fired power stations constructed in the 1960s. Existing power stations can be upgraded to use clean coal technology.

Conventional natural gas, which is generally found within a few thousand metres or so of the surface of the Earth, has accounted for most of the global supply to date. However, in recent years 'unconventional' deposits have begun to contribute more to supply. The main categories of unconventional natural gas are shown in Table 5.4.

Unconventional deposits are clearly more costly to extract but, as energy prices rise and technology advances, more and more of these deposits are attracting the interest of governments and energy companies.

Unconventional source of gas	Description
Deep gas	Deposits that exist very far underground (roughly 15,000 feet), well beyond 'conventional' drilling depths. Deep drilling has become more economical in recent years with improvements to exploration, and extraction techniques.
Tight gas	Gas that is trapped in unusually impermeable, hard rock, or in a sandstone or limestone in very tight formations that are impermeable and non-porous. Several techniques exist that allow natural gas to be extracted, including fracturing and acidising.
Gas-containing shales	Natural gas can exist in shale deposits (a very fine-grained sedimentary rock), formed 350 million years ago. The shale can be fractured to release the gas.
Coalbed methane	Many coal seams also contain natural gas (e.g. methane), either within the seam itself or the surrounding rock. What was once a by-product of the coal industry is becoming an increasingly important source of natural gas.
Geopressurised zones	Underground formations that are formed by layers of clay which are deposited and compacted very quickly on top of sand or silt (which are more porous and absorbent). Natural gas in this clay is squeezed out into the sand or silt by the rapid compression of the clay, under high pressure.
Arctic and sub-sea hydrates	These are made up of a matrix of frozen water that surrounds molecules of methane. Estimates suggest that methane hydrates may contain more organic carbon than the world's coal, oil and conventional natural gas combined.

▲ **Table 5.4** Unconventional sources of gas. Such techniques may be very important in buying time for more renewable energy to become available

China is the biggest consumer and producer of energy in the world. China accounted for 23 per cent of global energy consumption and 27 per cent of global energy consumption growth in 2016. The demand for energy in China continues to increase significantly as the country expands its industrial base. China's total energy consumption grew by 2.9 per cent in 2017.

The energy mix in China continues to evolve, and although coal remains the dominant fuel (Figure 5.10), accounting for 63.7 per cent of China's energy consumption in 2015, this figure is the lowest percentage share on record, down from highs of 74 per cent in the mid-2000s.

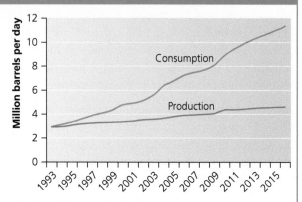

▲ **Figure 5.11** Chinese oil consumption and production, 1993–2016

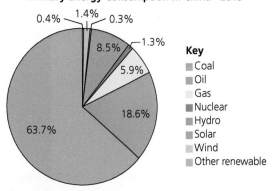

Primary Energy Consumption in China - 2015

Key
- ■ Coal
- ■ Oil
- ■ Gas
- ■ Nuclear
- ■ Hydro
- ■ Solar
- ■ Wind
- ■ Other renewable

▲ **Figure 5.10** Energy consumption in China by source, 2015

As the economy expanded rapidly in the 1980s and 1990s, much emphasis was placed on China's main energy resource, coal, in terms of both increasing production and building more coal-fired power stations. However, this was at the expense of huge environmental impact and an alarmingly high casualty rate among coal miners. According to Greenpeace, 80 per cent of China's carbon dioxide emissions comes from burning coal. China is the world's leading energy-related CO_2 emitter.

China was also an exporter of oil until the early 1990s, although it is now a very significant importer (Figure 5.11). China is the world's second largest consumer of oil, and moved from second largest net importer of oil to the largest in 2017 (Figure 5.12). This transformation has had a major impact on Chinese energy policy as the country has sought to secure overseas sources of supply. Long-term energy security is viewed as essential if the country is to maintain the pace of its industrial revolution.

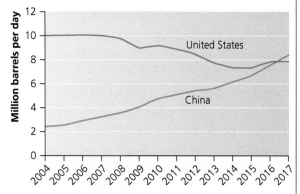

▲ **Figure 5.12** Annual USA and China gross crude oil imports, 2004–17. China overtook the USA in annual gross crude oil imports in 2017, importing 8.4 million barrels per day (b/d) compared with 7.9 million b/d for the United States

In recent years, China has tried to take a more balanced approach to energy supply and at the same time reduce its environmental impact. The Eleventh Five-Year Plan (2006–10) focused on two major energy-related objectives: (a) to reduce energy use per unit GDP by 20 per cent, and (b) to ensure a more secure supply of energy. Because of the dominant position of coal in China's energy mix, the development of clean coal technology is central to China's energy policy with regard to fossil fuels. China has emerged as the world's leading builder of more efficient, less polluting coal power plants. China has begun constructing such clean coal plants at a rate of one a month. The government has begun to require that power companies retire an older, more polluting power plant for each new one they build.

Heightened fears about oil supplies, energy security and climate change have brought nuclear power back onto the global energy agenda. The advantage of nuclear power over fossil fuels is that there are zero emissions of greenhouse gases. Increased usage of nuclear power means reduced reliance on imported fossil fuels (which can help ease concerns about energy security). Nuclear energy is not as vulnerable to fuel price fluctuations as oil and gas – uranium, the fuel for nuclear plants, is relatively plentiful and most of the main uranium mines are in politically stable countries.

The further development of nuclear power is an important strand of Chinese policy. By the end of 2009,

China had 11 operational nuclear reactors with a further 17 under construction. The World Nuclear Association (WNA) says that China has a further 124 nuclear reactors on the drawing board. Global nuclear power generation increased by 1.3 per cent in 2016, or 9.3 million tonnes of oil equivalent (mtoe), with China accounting for all of the net growth, expanding by 24.5 per cent (9.6 mtoe). Increased development of their nuclear energy capacity will lead to a reduction in carbon dioxide emissions and lead to a stabilisation or reduction of China's carbon footprint. Further rises in carbon dioxide emissions and global carbon cycle modifications on account of China's economic growth *are therefore not inevitable*.

② Anthropogenic and natural climate change

▶ *In what ways are human activities interlinked with global warming and how do anthropogenic impacts compare with natural causes of climate change?*

The greenhouse effect

Greenhouse gases are essential for life on Earth. The Moon is an airless planet that is almost the same distance from the Sun as is the Earth. Average temperatures on the Moon are about –18°C, compared with about 15°C on Earth. The Earth's atmosphere therefore raises temperatures by about 33°C. This is due to the greenhouse gases, such as water vapour, carbon dioxide, methane, ozone, nitrous oxides and chlorofluorocarbons (CFCs). They are called greenhouse gases because, as in a greenhouse, they allow short-wave radiation from the Sun to pass through them, but they trap outgoing long-wave radiation, thereby raising the temperature of the lower atmosphere (Figure 5.13). The greenhouse effect is both natural and good – without it there would be no life on Earth.

 KEY TERM

Greenhouse effect
Atmospheric gases absorb infra-red radiation, causing world temperatures to be warmer than they would otherwise be.

Increasing carbon emissions and the energy budget

The Earth's climate is driven by incoming short-wave solar radiation (Figure 5.13). A large proportion of the long-wave radiation emitted by the surface is absorbed by the atmosphere (clouds and greenhouse gases) from where it is re-radiated both back to the Earth's surface and to space.

Increasing carbon emissions (carbon dioxide and methane) mean that more heat is being radiated back towards the ground surface. The energy budget is changing; more heat is being retained, resulting in a warmer, more energetic climate system. However, there is great uncertainty even among experts over the extent and timing of future climate forecasts. This is known as the **enhanced greenhouse effect**.

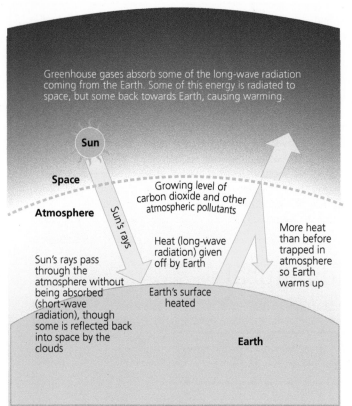

Greenhouse gases absorb some of the long-wave radiation coming from the Earth. Some of this energy is radiated to space, but some back towards Earth, causing warming.

Source: *OCR A2 Geography* by M. Raw (Philip Allan Updates, 2009) p.111

▲ **Figure 5.13** The greenhouse effect

Anthropogenic climate change

The enhanced greenhouse effect is built up of certain greenhouse gases as a result of human activity. Carbon dioxide (CO_2) levels have risen from about 315 ppm (parts per million) in 1950 to over 409 ppm in 2018 and are expected to reach 600 ppm by 2050. The increase is due to human activities: burning fossil fuels (coal, oil and natural gas) and deforestation. Deforestation of the tropical rainforest is a double blow – not only does it increase atmospheric CO_2 levels, it also removes the trees that convert CO_2 into oxygen. Deforestation accounts for about 12–15 per cent of global CO_2 emissions. The resulting 'hockey stick graph' of increasing carbon dioxide concentrations (see page 181, Figure 6.12) is a key piece of evidence for anthropogenic changes to the global carbon budget.

Methane (CH_4) is the second largest contributor to global warming, and is increasing at a rate of between 0.5 and 2 per cent per annum. Cattle alone give off between 65 and 85 million tonnes of methane per year. Natural wetland and paddy fields are another important source – paddy fields emit up to 150 million tonnes of methane annually. As global warming increases, bogs trapped in permafrost will melt and release vast quantities of methane (see pages 145–6, and 206).

> **⚷ KEY TERM**
>
> **Enhanced greenhouse effect** The process in which human activities have led to an increase in the amount of greenhouse gases in the atmosphere, and an increased trapping of greenhouse gases, leading to an increase in global temperatures, i.e. global warming.

> The greenhouse warming potential of methane is 25x that of CO_2 (at 100 year timescales).

As long as the amount of water vapour and carbon dioxide stay the same and the amount of solar energy remains the same, the temperature of the Earth should remain in equilibrium. However, human activities are upsetting the natural balance by increasing the amount of CO_2 in the atmosphere, as well as the other greenhouse gases.

Natural climate change

There are many causes of global warming and climate change. Natural causes include the following:

Drivers of climate change

- Greenhouse gases are produced by a range of natural phenomena such as:
  - volcanic activity
  - methane released by animals and peat bogs.
- Changes in the amount of dust in the atmosphere (partly due to volcanic activity), blocking out solar radiation.
- Milankovitch cycles (see also page 19, Figure 1.17):
  - Changes in Earth's tilt and variation in orbit around the Sun: every 100,000 years, the Earth's orbit changes from spherical to elliptical, changing the solar input.
  - Changes in the aspect of the poles from towards the Sun to away from it: the Earth's axis is tilted at 23.5°, but this changes over a 41,000-year cycle between 22° and 24.5°, also affecting solar input.
  - The Earth's axis wobbles, changing over 22,000 years, bringing further climate change.
- Natural fluctuations in atmospheric circulation (e.g. El Niño and La Niña) – see pages 16–18.
- Plate tectonic movements – such as Antarctica's arrival at the South Pole and the subsequent formation of the Antarctic ice sheet – have had an important influence on the Earth's climate on a timescale of many millions of years (see also page 18). Antarctica's arrival at the South Pole and the formation of the Antarctic ice sheet caused global sea level to fall by approximately 70 metres, reducing the size of the ocean carbon store.
- Changes in the Earth's ocean currents as a result of continental drift. The opening and closing of oceanic gateways between land masses may have altered global ocean circulation patterns, leading to climate changes.
- The uplift of mountain belts resulting from plate movements may have disrupted atmospheric circulation and triggered climatic changes.

> Natural forcings of climate change, such as the Milankovitch cycles, occur over tens of thousands of years, and have led to glacial and interglacial periods (page 19).

Feedback mechanisms

- Bush fires, caused by lightning strikes, releasing carbon dioxide into the atmosphere.
- Changes in temperature and changes in albedo due to position and extent of ice sheets. Increased ice cover leads to increased reflectivity of the Earth's surface and decreased average surface temperatures (see page 206).
- Changes in albedo due to variations in cloud cover (water is a greenhouse gas).

As you can see, the list of possible causes is a long one. It is interesting to reflect how the changes shown are driven by the operation of numerous different physical systems. As a good exercise, review the points above and identify synoptic links with different geography topics such as the water cycle, tectonic systems and different climatic subsystems.

Implications of climate change for physical systems and life on Earth

▶ *How are increased global temperatures impacting global biodiversity and feedback mechanisms relating to the carbon cycle?*

The effects of increased global temperature change

The effects of global warming are varied (see Table 5.5). Much depends on the scale of the changes. For example, some impacts could include:

- a rise in sea levels, causing flooding in low-lying areas such as the Netherlands, Egypt and Bangladesh – up to 200 million people could be displaced
- 200 million people at risk of being driven from their homes by flood or drought by 2050
- 4 million km² of land, home to one-twentieth of the world's population, threatened by floods from melting glaciers
- an increase in storm activity, such as more frequent and intense hurricanes (owing to more atmospheric energy)
- changes in agricultural patterns, for example a decline in the USA's grain belt, but an increase in Canada's growing season
- reduced rainfall over the USA, southern Europe and the Commonwealth of Independent States (CIS), leading to widespread drought
- 4 billion people could suffer from water shortages if temperatures rise by 2°C
- a 35 per cent drop in crop yields across Africa and the Middle East expected if temperatures rise by 3°C
- 200 million more people could be exposed to hunger if world temperatures rise by 2°C; 550 million if temperatures rise by 3°C
- 60 million more Africans could be exposed to malaria if world temperatures rise by 2°C
- extinction of up to 40 per cent of species of wildlife if temperatures rise by 2°C.

Positive effects	Negative effects
• An increase in timber yields (up to 25% by 2050), especially in the north (with perhaps some decrease in the south), resulting in increased carbon sequestration. • A northward shift of farming zones by about 200–300 km per 1°C of warming, or 50–80 km per decade, will improve some forms of agriculture, especially pastoral farming in the north-west. • Enhanced potential for tourism and recreation as a result of increased temperatures and reduced precipitation in the summer, especially in the south.	• Increased damage effects of increased storminess, flooding and erosion on natural and human resources and human resource assets in coastal areas. • An increase in pest species in some regions, especially insects, as a result of northward migration from the continent and a small decrease in the number of plant species due to the loss of northern and montane (mountain) types. • An increase in soil drought, soil erosion and the shrinkage of clay soils, leading to decreased plant growth and decreased carbon sequestration.

▲ **Table 5.5** Some potential effects of a changing climate in the UK, including carbon cycle impacts

The effects of climate change vary with the degree of temperature change (Figure 5.14; see also Chapter 1, pages 14–16, for information from IPCC reports). Climate change poses a threat to the world economy and it will be cheaper to address the problem than to deal with the consequences. The global-warming argument seemed a straight fight between the scientific case to act, and the economic case not to. Now, economists are urging action.

Effect on agriculture

Change in climate can lead to changes in weather patterns and rainfall (in both quantity and distribution). Climates may become more extreme and more unpredictable. An increase in more extreme weather conditions (e.g. hurricanes) can be expected as atmospheric patterns are disturbed.

Food	Falling crop yields in many areas, particularly LEDCs				
	Possible rising yields in some high latitude regions			Falling yields in many MEDCs	
Water	Small mountain glaciers disappear – water supplies threatened in several areas	Significant decreases in water availability in many areas, including Mediterranean and southern Africa		Sea level rise threatens major cities	
Ecosystems					
	Extensive damage to coral reefs	Rising number of species face extinction			
Extreme weather events	Rising intensity of storms, forest fires, droughts, flooding and heatwaves				
Risk of abrupt and major irreversible changes		Increasing risk of dangerous feedbacks and abrupt, large-scale shifts in the climate system			
0 °C	1 °C	2 °C	3 °C	4 °C	5 °C

Global temperature change (relative to pre-industrial)

▲ **Figure 5.14** Projected impacts of climate change

Agriculture will be affected. Drought reduces crop yield, and the reduction in water resources will make it increasingly difficult for farmers in many areas to irrigate fields.

Changes in the location of crop-growing areas can be expected, with movements north and south from the Equator: recent models predict dramatic changes to the wheat-growing regions of the USA, with many becoming unviable by 2050. This would have serious knock-on effects for the carbon cycle and the USA economy. Crop types may need to change and changing water resources will either limit or expand crop production depending on the region and local weather patterns.

Climate change, higher average temperatures and changing precipitation patterns may have three direct impacts on soil conditions. The higher temperatures cause higher decomposition rates of organic matter. Organic matter in soil is important as a source of nutrients and it improves moisture storage. More floods will cause more water erosion, while more droughts will cause more wind erosion.

Effect on biodiversity

Climate change may also lead to a reduction in biodiversity as species change their distribution in response to changes in climate. Some species – especially high altitude and high latitude species – have fewer options for migration, and so are more endangered.

Coral bleaching

Reef-building corals have a symbiotic relationship with a microscopic unicellular algae called zooxanthellae (Figure 5.15).

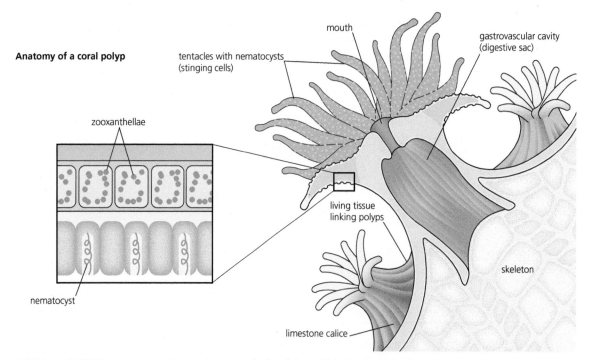

Anatomy of a coral polyp

zooxanthellae

nematocyst

tentacles with nematocysts (stinging cells)

mouth

gastrovascular cavity (digestive sac)

living tissue linking polyps

skeleton

limestone calice

▲ **Figure 5.15** The anatomy of a coral: zooxanthellae living within the polyp animal photosynthesise to produce food for themselves and the coral polyp, and in return are protected

The polyp animals secrete a protective skeleton of calcium carbonate, which forms the foundation of the coral reef ecosystem. The coral therefore forms an important store of carbon (Chapter 4). Reef-building corals need warm, clear water. Unfortunately, pollution, sedimentation, global climate change and several other natural and anthropogenic pressures threaten this fundamental, biological need, effectively halting photosynthesis of the zooxanthellae and resulting in the death of the living part of the coral reef.

Coral bleaching (Figure 5.16) can be caused by increases in water temperatures of as little as 1–2°C above the average annual maxima.

In 1998 there was extensive and intensive bleaching that affected the majority of coral reefs around Puerto Rico and the northern Caribbean. In the south-west region, a large number of coral colonies bleached completely (100 per cent of the living surface area) down to 40 metres deep. Maximum temperatures measured during 1998 in several reef localities ranged from 30.15°C (20 metres deep) to 31.78°C at the surface.

▲ **Figure 5.16** When environmental conditions become stressful, zooxanthellae may leave the coral, leaving it in an energy deficit and without colour – a process that is referred to as coral bleaching. If the coral is re-colonised by zooxanthellae within a certain time, the coral may recover, but if not the coral will die

Temperatures have risen in all oceans in the last 40 years, as seen from satellite images and other measures over 135 years from the National Oceanic and Atmospheric Administration of the USA. This means that coral bleaching will remain a major cause of global biodiversity loss and impact the global carbon cycle by decreasing the amount of carbon stored in coral and increasing the amount in the atmosphere.

Methane feedback

Around one-quarter of the Earth's surface is affected by continuous or sporadic permafrost, including tundra, polar and mountain regions. Globally, permafrost covers 23 million square kilometres (mostly in Earth's northern hemisphere). It formed during past cold glacial periods and has persisted through warmer interglacial periods, including the Holocene (the last 10,000 years).

Reports indicate that methane-storing permafrost is now shrinking at an alarming rate. According to scientists at NASA, temperatures in Newtok, Alaska, have risen by 4°C since the 1960s, and by as much as 10°C in winter months. The effects of permafrost melting are potentially magnified over time by positive feedback loops.

1 As the atmosphere warms, more permafrost is expected to melt.
2 This will release large amounts of methane (some researchers estimate that the volume of this gas stored in permafrost equates to more than double the amount of carbon currently in the atmosphere).
3 The atmosphere will warm up even more quickly.
4 Even more methane will be released by more melting permafrost, etc.

In theory, this could easily take Earth's climate beyond a 'tipping point' (an irreversible move away from one state of equilibrium to another).

Terrestrial and marine carbon feedback

Figure 5.17 shows another positive feedback model. Here, combined terrestrial and marine feedback loops are shown which both result from, and accelerate further, global temperature rise. The elements of this model include:

- increasing water vapour in the atmosphere; because water vapour is also a greenhouse gas, this could lead to further temperature rises (however, huge scientific uncertainty exists in defining the extent and importance of this positive feedback loop: as water vapour increases in the atmosphere, more of it will eventually also condense into clouds, which are more able to reflect incoming solar radiation)
- terrestrial permafrost melting and methane release
- a reduction in seawater's ability to absorb surplus CO_2 from the atmosphere: warmer water is less effective at absorbing CO_2 than colder water, and so may begin to release, rather than absorb, the gas if temperatures continue to rise.

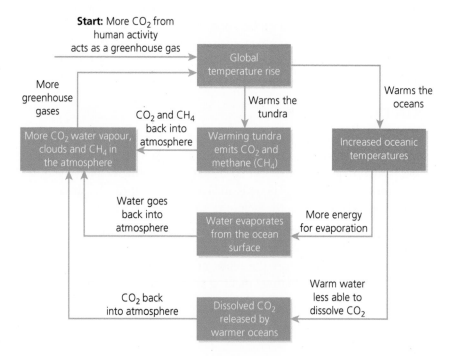

▶ **Figure 5.17** A system diagram showing how interrelations between climate change, water flows and carbon flows could create positive feedback loops

There are further terrestrial and marine feedback loops to consider:

- Ocean acidification could impact negatively on coral and marine ecosystem health in ways that reduce biological sequestration of CO_2 in the oceans, i.e. the biological pump might become less effective (see page 108).
- The global biome pattern may change in response to rising Global Mean Surface Temperature (GMST). If the tree line moves north – and the coniferous tree biome grows in size – then more carbon might be stored. But if large areas of grassland are damaged by desertification, carbon storage will be lost. Overall, there is huge uncertainty over what the net effect of higher temperature rises will be on carbon storage patterns linked with biome distributions.
- Increased carbon dioxide in the atmosphere may have a 'fertilisation effect', leading to increased plant productivity. Increased plant growth would reduce carbon dioxide levels in the atmosphere – a negative feedback loop.

ANALYSIS AND INTERPRETATION

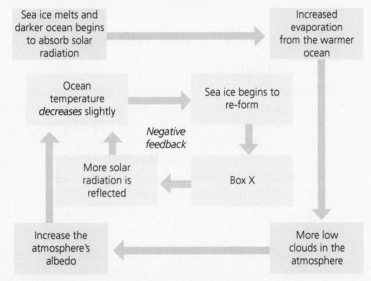

▲ **Figure 5.18** A prediction of how negative feedback could help reduce future cryosphere shrinkage on account of a warming atmosphere

(a) Identify the change shown in Box X.

GUIDANCE

This is a straightforward question asking to use your knowledge of system feedback to identify a process involved in the system indicated.

(b) Using Figure 5.18 and your own knowledge, assess the importance of negative feedback processes for supporting life on Earth.

GUIDANCE

Note that this question addresses negative rather than positive feedback (the latter is usually associated with climate change through anthropomorphic emissions of carbon). It shows how negative feedback could help the climate system to self-correct if Arctic ice begins to melt. Think about how the control of the Earth's ice cover and temperature, indicated in the figure, may assist life on Earth.

(c) Suggest why the negative feedback prediction shown in Figure 5.18 might fail to operate if anthropogenic carbon emissions remain at a high level.

GUIDANCE

This question asks you to apply your knowledge of anthropogenically induced climate change to predict possible impacts on the model shown in Figure 5.18. Think about the concepts of equilibrium and tipping points to inform your answer, and the time periods over which negative and positive feedback mechanisms operate.

The implications of system feedback for human societies

There are also serious implications linked with the potential disappearance of large parts of the cryosphere (see pages 19–20), the meltwater of which currently feeds large rivers upon which hundreds of millions of people depend for their water security. For example, China is home to 1.3 billion people, many of whom have recently moved to urban areas. By 2025, there will be 220 Chinese cities with populations in excess of one million, eight of which will be megacities (settlements with over 10 million residents). However, the sustainability of these growing settlements is threatened by a new lack of water security for the south-east Asian region as a whole.

- Many major rivers are fed by seasonal meltwater runoff from major glaciers in the region, notably the Himalayan Plateau.
- Every summer, ice melts and feeds Asia's largest rivers. Fresh snowfall each winter replenishes the glaciers, meaning that over time the meltwater cycle is sustainable.
- However, climate change system feedback threatens to permanently reduce the size of glacial ice stores in the region. Although this will increase meltwater in the short term, in the long term it could lead to dangerous water shortages (see Chapter 3, pages 69–75).

Evaluating the issue

▶ *To what extent have human impacts on the carbon cycle led to a new geological epoch, the Anthropocene?*

Identifying possible contexts and criteria for the evaluation

The focus of this chapter's critical evaluation is the Anthropocene concept. This idea appeared previously in Chapter 1, where we learned that a growing number of scientists believe Earth history has entered a new Quaternary period epoch. They argue that the Holocene has now given way to the Anthropocene – a new era characterised by widespread changes to global and local physical systems alike on account of the way there are now 'human fingerprints' all over almost everything in the natural world. In particular, it is the unprecedented rapid change in atmospheric carbon dioxide witnessed since the onset of the industrial revolution – and resulting carbon cycle changes – that underpins the Anthropocene argument.

- For most of the history of Earth, the planet has been subject to natural forces, such as water and carbon cycling, which have influenced and shaped the life that inhabits it. Since the advent of modern humans, one species – *Homo sapiens* – has been the dominant influence on Earth's ecosystems.
- Early humans lived within natural Earth systems, as hunter-gatherers, and had little impact on their environment. Populations were low in numbers and people lived off the land, rather than manipulating it towards their own ends. As humanity spread from Africa approximately 120,000 years ago, becoming farmers and clearing land to grow crops, the impact of *Homo sapiens* on the planet became more severe. The development of settled agriculture represents one of the

most significant changes in human history, enabling human populations to start growing. This period, known as the Neolithic ('new stone age') revolution, began in the 'fertile crescent' in the Middle East about 10,000 years ago, and changed forever the way that humanity interacts with the environment. The growth in agriculture led to land clearance and a significant transformation of the terrestrial biosphere.

But to what extent are changes in the carbon cycle and other physical systems during recent decades representative of *fundamental* changes in environmental conditions, as opposed to fluctuations occurring within a broadly steady state (see page 10) when viewed in the long term? The answer may depend on the extent to which we believe changes are reversible or not (and whether certain system thresholds have been crossed or tipping points reached).

Possible contexts that can be used to argue the case for and against the Anthropocene thesis include:

- The historical evidence showing changing concentrations of carbon dioxide over different timescales ranging from hundreds to millions of years. It is important to look not only at changes in the total size of the atmospheric carbon store; additionally, consideration should be given to the timescale of changing carbon fluxes – is there a historical precedent for the so-called 'hockey stick' trend in greenhouse gases over the last century, for example?
- Changes in the way the ocean operates as a carbon store and the implications of acidification for sea life: because the oceans cover two-thirds of the Earth's surface, any

evidence suggesting they are entering a new system state could be viewed as compelling evidence that the Anthropocene has arrived.

- Biodiversity patterns and trends and the extent to which humans are responsible for a mass extinction that is compatible with past 'boundary events' (for example at the end of previous geological epochs such as the Cretaceous period). Changes in biodiversity may have far-reaching consequences for carbon cycling, for example when large areas of forest were placed with monoculture: this has a significant impact on seasonal and long-term carbon sequestration.

Evaluating the view that we have entered a new geological epoch

For the majority of the time since *Homo sapiens* evolved in Africa, about 250,000 years ago, the concentration of CO_2 in the atmosphere was low (below 200 ppm). The concentration of carbon dioxide rose for centuries at a time (but not above 240 ppm) and then fell for longer periods of time. This pattern steadied at 240 ppm from the period demarcated by the beginning of agriculture (Figure 5.19). Early human activities that may have contributed to relatively small elevated levels of CO_2 (with evidence provided by ice cores) included fire-stick farming and forest clearing.

In recent times, humanity's tremendous growth in population, from mere millions in the Neolithic period to over 7 billion today, and its tremendous thirst for resources and land, has put even more pressure on the Earth's natural systems. The discovery of coal and other fossil fuels as a source of energy liberated humanity from the shackles of horse-drawn equipment to farm the land, and enabled much larger areas to be cleared and used for crops. Energy – along with carbon – trapped by plants millions of years ago could be released: the amount of energy available to humans increased hugely. Since the start of the industrial revolution in Europe, in the early part of the nineteenth century, CO_2 has been released at an increasing rate (Table 5.6), attributed to the burning of coal and oil. These 'fossil fuels' were mostly laid down in the Carboniferous period. As a result, we are now adding to our atmosphere carbon that had been locked away for about 350 million years. This is an entirely new development in geological history.

	CO_2/ppm
pre-industrial revolution level	280 (±10)
by mid-1970s	330
by 1990	360
by 2007	380
by 2013	400
by 2018	409
by 2050 (if current rate is maintained)	500

▲ **Table 5.6** Changing levels of atmospheric CO_2 – recorded and predicted

The stages of the Anthropocene
Pre-Anthopocene events:
fire-stick farming, mega fauna extinctions, forest clearing
Anthopocene Stage 1 (c.1800–1945):
internal combustions engine, fossil fuel energy, science and technology
Anthopocene Stage 2 (c.1945–2010 or 2020):
the Great Acceleration, new institutions and vast global networks
Anthropocene Stage 3 (2010 or 2020?):
sustainability or collapse?

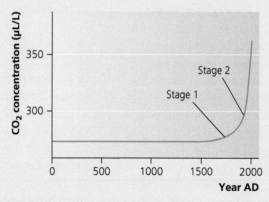

▲ **Figure 5.19** The stages of the 'Anthropocene' NB: The anthropocene is not only about climate change: this is one large way in which humans have altered the earth system but not the only one

To assess whether the current period should be attributed its own epoch, the geological past may hold clues for similar inputs of carbon into the atmosphere. The Palaeocene-Eocene Thermal Maximum (PETM), which occurred 56 million years ago (Figure 5.20), involved more than 5°C of warming over a period of 15–20 thousand years. The warming was caused by an input of more than 2000 gigatonnes of carbon into the atmosphere (1 gigatonne = 1 billion tons). Over the next few centuries, if emissions of anthropogenic carbon dioxide (CO_2) continue largely uninterrupted, a total of 5000 PgC (petagram of carbon, where 1 petagram = 10^{15}g) may enter the atmosphere, causing global temperatures to increase by more than 8°C and surface ocean pH to decrease by roughly 0.7 units. This degree of carbon release will not have been seen since the PETM. During the PETM, an initial carbon pulse of about 3000 PgC took place over approximately 6000 years.

One critical difference between the Anthropocene and PETM carbon release events is the timescale of carbon input (Figure 5.21). While it is clear that the carbon input during the PETM was rapid on geological timescales (a few thousand years), anthropogenic inputs have been much more rapid. Studies suggest a PETM release of carbon at a modest 0.2 gigatonnes per year (peaking at 0.58 gigatonnes), compared to a much larger roughly 10 gigatonnes of carbon being added by humans to the atmosphere each year.

The oceans have absorbed approximately one-third of the CO_2 emitted by humans, over the period from 1750 to 2000; this absorption has already caused a decrease in surface ocean pH by approximately 0.1 units from roughly 8.2 to 8.1. This acidification of the oceans will have a negative effect on sea life (see page 147). On relatively long timescales, greater than 10,000 years, feedback mechanisms involving weathering and burial of calcium carbonates regulate the saturation of carbon in the oceans and mitigate ocean acidification. Over shorter

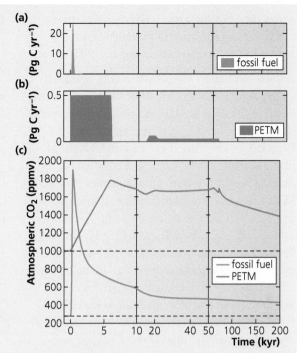

▲ **Figure 5.20** Long-term legacy of massive carbon input to the Earth system: Anthropocene versus Palaeocene–Eocene Thermal Maximum (PETM, approx. 56 Ma). (**a**) Fossil fuel emissions: total of 5000 PgC over 500 years, and ongoing; (**b**) PETM carbon release: 3000 PgC over 6 k^{yr} plus approximately 1500 Pg over more than 50 k^{yr}; (**c**) Simulated evolution of atmospheric CO_2 in response to the carbon input using a computer model (LOSCAR: Long-term Ocean-atmosphere-Sediment CArbon cycle Reservoir). The LOSCAR model is designed to efficiently compute the partitioning of carbon between ocean, atmosphere, and sediments on timescales ranging from centuries to millions of years. Note different y-axis scales in graphs (**a**) and (**b**)

time periods (i.e. few hundred years), such as those represented by current CO_2 release, rapidly increasing CO_2 levels cannot be mitigated by feedback as they are too slow and do not operate on timescales of decades to centuries. The timescale of the anthropogenic carbon input is therefore too short, compared to the PETM, for the natural capacity of the ocean stores to absorb the additional carbon at the rate it is being generated. Research suggests, therefore, that ocean acidification and its effects on marine calcifying organisms will probably be more severe in the future than during the PETM.

A mass extinction is a period in which at least 75 per cent of the total number of species on the Earth at the time are wiped out. The fossil record shows that, over millions of years, there have been five mass extinctions, caused by natural physical (abiotic) phenomena. These events include the Ordovician–Silurian extinction (439 Ma, where Ma is used to indicate 'millions of years ago'), Late Devonian (364 Ma, Permian–Triassic (251 Ma), End-Triassic (199 Ma) and Cretaceous–Tertiary extinction (65 Ma). In mass extinctions, species disappear in a geologically short time period, usually between a few hundred thousand to a few million years. The fossil record indicates that recovery of biological diversity generally takes several million years after a mass extinction event.

Studies suggest that a sixth mass extinction caused by humans will more likely resemble the Permian–Triassic and Cretaceous–Tertiary extinctions than those that occurred during the PETM: analysis of the marine fossil record suggests that, this being the case, recovery will take tens of millions of years.

The Quaternary period includes two epochs, the Pleistocene and the Holocene. The latter began about 11,500 years ago, and is signified by an interglacial period. The Holocene is, however, just one of a series of warmer periods that have occurred over the last 2 million years, and is defined as an epoch simply because many of the soils and river deposits on which we live were formed during this time. Many would argue that the modification of biomes by humans over the recent past has led to some people terming them 'anthromes' (i.e. human-modified biomes – Ellis 2015), and other changes to the land, sea and air, including mass extinction events, warrants the addition of a third epoch to the Quaternary. The concept of the Anthropocene seems justified (see page 20).

Evaluating the view that we have *not* entered a new geological epoch

Although it is clear that since the industrial revolution the amount of carbon stored in the atmosphere has risen from around 280 ppm to its current figure of 409 ppm, fluctuations in carbon dioxide concentrations in the atmosphere, including very large injections of the gas, is nothing new in terms of planetary history, nor is global warming as Earth's climate has changed many times in the past.

Geologists believe that very cold ice ages occurred around 650–750 million years ago, when much of the Earth was covered with ice and snow: the cryosphere was much larger than today. Global patterns of vegetation and soil carbon storage would have been vastly different, and large areas of the planet which currently support vegetation would have supported none. It is possible that more carbon was stored in the oceans during colder climatic periods in the past because the amount of carbon stored in the ocean water in a dissolved form increases when water temperatures are lower.

The Earth has also experienced very hot conditions. Around 30–60 million years ago the poles were probably ice-free and sea levels would have been much higher. Therefore, ocean storage of carbon might have been higher on account of the greater volume of water, although this might have been offset by warmer temperatures, meaning that the amount of carbon stored per litre of water would have been less.

The important role of non-human factors
Another view is that – when a long-term view is taken – natural forces, rather than human factors, have a more important influence on Earth's history and the arrival of different geological epochs.

- For example, plate tectonic movement is an important and ongoing natural cause of changing carbon storage (albeit over a long timescale).

- Antarctica's arrival at the South Pole and the formation of the Antarctic ice sheet caused global sea level to fall by an estimated 70 metres, thereby reducing the size of the ocean carbon store.
- The uplift of the Tibetan Plateau and the Himalayas is another important geological episode that affected carbon storage. These mountains are composed of limestone and were originally sea floor deposits that have been thrust upwards into the atmosphere by the convergence of two tectonic plates. The subsequent erosion of these mountains – and the removal of carbon in solution after carbonation has taken place – has reduced the amount of carbon stored there.

Arriving at an evidenced conclusion

Industrialisation and globalisation have caused an unprecedented increase in atmospheric carbon storage with many knock-on effects for ecosystems and oceans, warranting the use of the term 'Anthropocene' to reflect this fundamental change in Earth systems. The Anthropocene can be seen as a natural result of the evolution of a tool-using intelligent species, humankind, and the consequent rise of technological civilisation (Allenby, 2015). Changes in the carbon cycle and climate can be seen as fundamental outcomes of the evolutionary advantage acquired by humankind through its use of geological supplies of energy.

If current trends in carbon emissions continue, humans will (assuming fossil fuel reserves are sufficient) release several thousand Pg of carbon into the atmosphere, with severe consequences for the planet (see pages 145–6, regarding positive feedback mechanisms), leading to the extinction of species on a mass scale. The rate of change of carbon emissions and temperature change is more rapid than anything that has been seen in Earth's past, far and above natural causes of climate change. This suggests that the period in which we live does indeed warrant being termed the 'Anthropocene'. In order to limit the total carbon input to 1000 PgC and stretch emissions over 500 years, global carbon emissions would need to be cut in half over the next 30 years, starting tomorrow. If humanity is able to do this, perhaps the jury is still out about whether we are living through a unique geological epoch – the evidence will be, should we fail, all too evident for our descendants to see.

 KEY TERM

Fire-stick farming A technique used by Indigenous Australians to help hunt animals. Land was ignited and the resulting fires herded animals into specific areas where they could be killed. The technique also caused new vegetation to grow which attracted new animals.

Chapter summary

✔ Carbon-emitting energy resources are at the core of the global economy. The energy mix of a country is comprised of domestic and foreign, renewable and non-renewable. Access to and consumption of energy resources depends on physical availability, cost, technology, public perception, level of economic development and environmental priorities. Countries are largely dependent on fossil fuels to drive economic development.

✔ Energy security is the uninterrupted availability of energy sources at an affordable price. Energy pathways (pipelines, transmission lines, shipping routes, road and rail) are a key aspect of security but can be prone to disruption, especially as conventional fossil fuel sources deplete. There is a mismatch between locations of conventional fossil fuel supply (oil, gas, coal) and regions where demand and carbon emissions are highest, resulting from physical geography.

✔ While human activity is almost certainly responsible for the current phase of global warming, natural changes in the carbon cycle (and thus global temperatures) also occur over time, ranging from volcanic activity to changes in Earth's orbit.

✔ Emissions of greenhouse gases from non-renewable energy use are the main contributors to human-induced climate change. The combustion of fossil fuels has altered the way in which energy from the Sun interacts with the atmosphere and the surface of our planet.

✔ The concentration of atmospheric carbon (carbon dioxide and methane), and associated positive feedback mechanisms, strongly influences the natural greenhouse effect, which in turn determines the distribution of temperature and precipitation. Changes to the carbon budget have an impact on land, ocean and atmosphere, with fundamental implications for life on Earth.

Refresher questions

1 Using evidence in your answer, outline the main global trends and patterns in (i) fossil fuel use, and (ii) energy security.

2 What is meant by 'peak oil' and why is it important for future energy security?

3 Using examples, outline factors which affect a country's choice of energy generation.

4 Explain the strengths and weaknesses of nuclear power use as an alternative to fossil fuel burning.

5 Using examples, explain the difference between anthropogenic and natural climate change.

6 Draw an annotated diagram to explain the natural greenhouse effect. Outline what is meant by the enhanced greenhouse effect.

7 Outline the projected effects of anthropogenic climate change on the structure and functioning of Earth's carbon cycle.

8 Explain why climate change is harmful for coral reefs.

9 Using the example of Earth's carbon cycle, explain the Anthropocene concept.

Discussion activities

1 In a small group, prepare a presentation explaining the greenhouse effect and its importance for sustaining life on Earth. Go on to explain the enhanced greenhouse effect and its causes.

2 Organise a class debate about the view that current global warming could be a natural and not human-induced phenomenon.

3 In pairs, discuss the main factors that hinder the development of sustainable energy strategies with a lower carbon footprint than fossil fuels.

4 In small groups, outline possible actions governments could take to reduce national and thus global carbon emissions.

5 Discuss the value of the system theory in Chapter 1 of this book in relation to (i) understanding why Earth's carbon cycle and climate are changing, and (ii) our ability to make predictions about what will happen if insufficient action is taken to reduce humanity's carbon footprint.

FIELDWORK FOCUS

A *Profiling the carbon footprint of a local place, community or institution.* Many research tools exist that can help you assess the impact your community is having on the global carbon cycle, using calculators such as http://footprint.wwf.org.uk/. A range of secondary and primary data, including interviews, could be used to document what particular strategies individuals may be using to limit their personal footprints. As part of the profile, local carbon footprint(s) could be compared with other places (using secondary data); or a critical assessment could be made of the value of any mitigation measures in terms of their overall significance.

B *Evaluating evidence for the onset of the Anthropocene in a local area.* This could be a challenging but potentially rewarding investigation focus with many opportunities for a rigorous evaluation to be carried out afterwards that reflects critically on different aspects of the work. In the first place, we would need to devise a programme of data collection that is fit for purpose. What criteria might this include? Possibilities include: changes in local climate (e.g. when spring begins); vegetation (the kinds of plants that grow and thrive in people's gardens and parks); changes in local river flows or other environmental features. A mixture of secondary and primary data sources can be used, including old photographs and paintings (as proxy data for what the local environment used to be like). Interviews with older community members might focus on people's recollections about the climate (how much it used to snow, when autumn began, etc.). A similar approach can be taken if you decide instead to evaluate evidence for changing carbon cycles in your local area.

Further reading

Allenby, B. (2015) 'Climate Redux: Welcome to the Anthropocene', *Issues in Science and Technology* 31 (3), pages 37–9, University of Texas at Dallas

Ellis, E.C. (2015) 'Ecology in an Anthropogenic Biosphere', *Ecological Monographs* 85 (3), pages 287–331, published by Wiley on behalf of the Ecological Society of America

Raupach, M.R. and Canadell, J.G. (2010) 'Carbon and the Anthropocene', *Current Opinion in Environmental Sustainability* 2, pages 210–18, Elsevier, doi:10.1016/j.cosust.2010.04.003

Ruppel, C. (2006) 'The G Forces of Energy Insecurity', www.resilience.org/stories/2006-06-10/g-forces-energy-insecurity

Stern, N. (2006) *Stern Review on the Economics of Climate Change,* HM Treasury, London

Taylor, M. (2014) *Global Warming and Climate Change,* Chapter 2: Loading the dice: humans as planetary force, ANU Press, www.jstor.org/stable/j.ctt130h8d6.7

Zeebe, R.E and Zachos, J.C. (2013) 'Long-term Legacy of Massive Carbon Input to the Earth System: Anthropocene versus Eocene', *Philosophical Transactions of the Royal Society* A, 371: 20120006, http://dx.doi.org/10.1098/rsta.2012.0006

CHAPTER 6

Climate change mitigation and adaptation

Management strategies can be implemented, both locally and globally, to protect the carbon cycle as regulator of the Earth's climate. These strategies include greater use of renewable energy and reducing emissions through carbon capture and storage. Mitigation addresses the causes of climate change and operates to reduce anthropogenic effects. Adaptation makes changes to nullify the effects of climate change, thereby building up the resilience of communities and ecosystems to climate change. This chapter:

- explores the decarbonisation of economic activity and development of renewable energy
- analyses the strengths and weaknesses of carbon capture and storage
- investigates climate change adaption and resilience
- evaluates the extent to which a warmer world is inevitable.

KEY CONCEPTS

Adaptation The ability to respond to changing events and to reduce current and future vulnerability to change. Societies must adapt to climate change, for instance.

Mitigation The reduction of a phenomenon that is having a negative effect on people, places or the environment. Climate change mitigation is an attempt to reduce greenhouse gas emissions (by an individual or society).

Resilience The ability of an environment or society to adapt to changes that have a negative impact upon them, such as climate change.

Global governance This describes the steering rules, norms, codes and regulations used to regulate human activity at an international level. At this scale, regulation and laws can be tough to enforce, however.

 ## The decarbonisation of economic activity and renewable energy

▶ *How can world economies become less dependent on fossil fuel technologies and what role does renewable energy play in reducing the carbon footprint of energy generation?*

Global governance and the changing carbon cycle

Global, national and local management strategies can be employed either to reduce the effects of anthropogenic climate change or to adapt to changes brought about by it. Management strategies that aim to protect the functioning of the carbon cycle as regulator of the Earth's climate are known as mitigation strategies. These are actions that aim to (i) reduce and/or stabilise greenhouse gas (GHG) emissions, and (ii) remove from the atmospheric store those surplus GHGs that have been added by human activity since the start of the industrial revolution.

United Nations actions

As part of global governance, the international community has agreed a long-term goal to limit the increase in average global temperatures to no more than 2 degrees above the pre-industrial mean temperatures and to aim to limit the increase to 1.5°C. This was decided at the United Nations Climate Conference (COP21) in Paris, December 2015, where 195 countries adopted the first universal, legally binding global climate deal. In addition, governments agreed to strengthen society's ability to deal with the impacts of climate change and provide continued and enhanced international support for adaptation to developing countries.

Rebalancing the carbon cycle could be achieved through mitigation but this requires global scale agreement and national actions, both of which have proved to be problematic. There has been a long history of global attempts to limit carbon emissions (Figure 6.1). In 1992, at the Rio de Janeiro Earth Summit, the world's governments adopted the UN Framework Convention on Climate Change (UNFCCC). Its main objective is to 'achieve, in accordance with the relevant provisions of the Convention, stabilisation of greenhouse gas concentrations in the atmosphere at a level that would prevent dangerous anthropogenic interference with the climate system'. UNFCCC came into effect in 1994 but failed in its attempt to slow down greenhouse gas emissions.

The Kyoto Protocol, signed in 1997, was the first major attempt to implement the treaty. It gave all developing countries legally binding targets for cuts in emissions from the 1990 level by 2008–12. The EU agreed to cut emissions by 8 per cent, Japan by 7 per cent and the USA by 6 per cent. The outcomes of such international meetings have, in general, not achieved the high expectations they aspired to. Some countries found it easier to make cuts than others. Since 1992, when negotiations for the Kyoto Protocol first began, greenhouse gas emissions have risen by 50 per cent. In 2010, despite a global recession, they rose by 5 per cent.

Pearson Edexcel · AQA · OCR

KEY TERM

Mitigation The reduction of a phenomenon that is having a negative effect on people, places or the environment. Climate change mitigation is an attempt to reduce greenhouse gas emissions (by an individual or society).

▲ **Figure 6.1** The chronology of attempts to tackle global warming

The following text is transcribed from Figure 6.1:

IPCC produces its biggest report yet, stating there is a 90 per cent certainty that much of the observed warming of the climate can be attributed to human actions. George W. Bush changes his position, agreeing for the first time to enter international negotiations on a successor to the Kyoto Protocol. At Bali in December world governments agree to start two years of negotiations aimed at forging a successor to Kyoto

Durban Conference 2011 attempts to find binding cuts for LEDCs and MEDCs by 2015

2015 The Paris Conference seeks to limit temperature increase to 2°C above pre-industrial levels. Countries have to reduce their carbon use as soon as possible.

IPCC publishes its first report, finding that gases such as CO₂ increase the natural greenhouse effect

President Trump pulls the USA out of the Paris accord in 2017

The Copenhagen conference was to have marked the end of negotiations on a successor to Kyoto. But politicians said that a legally-binding agreement will not be signed until 2010

US President George W. Bush delivers a speech in which he rejects the Kyoto Protocol and casts doubt on the science behind climate change

UN Environment Programme and the World Meteorological Organisation set up Intergovernmental Panel on Climate Change (IPCC)

Former US Vice-President Al Gore releases *An Inconvenient Truth*, a documentary about climate change that becomes a worldwide hit. The film goes on to win best documentary Oscar in 2007 and he shares that year's Nobel peace prize with the IPCC

Earth Summit in Rio de Janeiro produces the United Nations Framework Convention on Climate Change (UNFCCC), which binds governments to take action to avoid dangerous climate change

Russia agrees to ratify the Kyoto Protocol. The move guarantees that the treaty will come into force

Kyoto Protocol signed. This protocol to the UNFCCC treaty sets out targets and deadlines by which developed countries must cut carbon emissions. Over the next few years, the protocol is ratified by all developed countries except the US and Australia

The European Union's emissions trading scheme comes into effect

Scientists attribute much of the steep rise in emissions of the previous few years to China's rapid economic expansion

Axis labels: World CO₂ (energy-related) emissions, Gigatonnes; Year; Forecast

Carbon mitigation strategies

Mitigation strategies reduce the causes of climate change by reducing the additional inputs of carbon into the atmospheric store that humans are responsible for. Such management strategies include the following:

- Afforestation: planting trees to establish where there were no forests previously (see page 171).
- Carbon taxation: these taxes raise the costs of coal, oil and gas compared to renewable sources of energy, shifting energy use towards the low-carbon options. These environmental taxes can be implemented by taxing the burning of fossil fuels (coal, petroleum products such as

gasoline and aviation fuel, and natural gas) in proportion to their carbon content.

- Energy efficiency: attempts to improve products and services so that less energy is required for them to function. For example, fluorescent lights and LED lights reduce the amount of energy needed to provide light for an area. Many modern appliances, such as fridges, freezers and washing machines, require much less energy to function compared with older models.

- Energy conservation: efforts to reduce energy consumption. For example, having homes with south-facing windows (in the northern hemisphere) allows greater solar heating of rooms. Other methods of conservation include double- or triple-glazing of windows, cavity wall insulation and loft insulation.

- Wetland restoration: recreating rich peatlands that were lost during agricultural development. For example, from 1640 onwards, large areas of wetland, such as the East Anglian fens in the UK, were drained for farming, which degraded the peat. No longer waterlogged, the peat shrank, decomposed and became eroded by the wind, which increased the flux of CO_2 to the atmosphere. Restoring these wetlands increases the land carbon store and decreases the atmospheric store. Degraded and drained peatlands emit more than 2 Gt CO_2e annually (gigatonne of carbon dioxide equivalent (GtC) is the unit used by the United Nations climate change panel, the Intergovernmental Panel on Climate Change (IPCC), to measure the amount of carbon in various stores – see Chapter 5, page 130): through rewetting of peatlands it is possible to restore carbon levels in peat soils that have already been degraded.

- Improving agricultural practices: agroforestry combines agriculture with forestry, allowing the farmer to continue cropping while using trees for fodder, fuel and building timber: trees protect, shade and fertilise the soil, decreasing rates of decomposition and related rates of respiration and increasing photosynthesis. Such practices improve carbon sequestration in agricultural soils and above-ground biomass through a range of soil, crop and livestock management practices, and protect existing carbon in the system by slowing decomposition of organic matter and reducing erosion. Other practices reduce the frequency with which the soils are tilled, and control erosion (e.g. terracing, contour strips and cover crops).

- Carbon capture and storage: when fossil fuels are burned, the carbon dioxide enters the atmosphere, where it may reside for decades or even centuries. A potential solution is to capture the carbon dioxide instead of allowing it to accumulate in the atmosphere. Two main ways to do this have been proposed. The first is to capture the gas at the site where it is produced (e.g. the power plant), and then to store it underground in a geologic deposit (e.g. an abandoned oil reservoir). The second is to allow the gas to enter the atmosphere but then to

 KEY TERM

Carbon capture The process of capturing carbon dioxide and depositing it where it will not enter the atmosphere.

remove it directly from the atmosphere using specially designed removal processes (e.g. collecting the carbon dioxide with special chemical sorbents that attract it). This latter approach is called direct air capture of carbon dioxide.

- Carbon market: a proposed way of using market forces to bring about carbon emissions reductions. Activities which help to store carbon – such as peatland restoration – might be rewarded with credits that can be offset against activities that produce CO_2.

- Reducing emissions through such methods as carbon trading and international agreements: carbon trading is an attempt to manage the amount of carbon dioxide released by different sectors/industries; this places a limit on total trading carbon offset schemes, in an attempt to reduce the overall impact of carbon emissions by investing in projects that cut emissions elsewhere.

- Reducing carbon emissions through utilisation of renewable sources of energy: renewable energy sources include solar, hydro-electric, geothermal, biomass and tidal schemes. These release either no direct sources of carbon dioxide during energy generation (solar, wind, tidal and hydro-electric) or can be considered carbon neutral, i.e. biomass uses plant material which has absorbed carbon dioxide, and so CO_2 released from biofuel combustion approximately equals the amount of CO_2 sequestered in biomass. Biomass can be a sustainable source of fuel as long as the annual use of biomass does not exceed the annual production of biomass for fuel.

KEY TERM

Renewable energy The energy from a source that is not depleted when used, such as wind or solar power.

Mitigation can be practised by different players (stakeholders) at different scales, from a citizen switching off a light, to a government setting strict national targets for reduced carbon emissions.

Given the scale of the issue, further planetary warming is inevitable (although see discussions in the final section of this chapter, pages 179–84), even if mitigation is highly successful. In addition, there is a time-lag between emissions and warming and it will take some considerable time at the global scale to bring carbon emissions under control. Even if mitigation strategies drastically reduce future emissions of greenhouse gases, past emissions will continue to have an effect for some time. Other management strategies can be employed to limit the impacts of climate change (adaptation strategies – see pages 175–6).

There is a limit, however, to how much humanity can adapt, and so mitigation is ultimately essential. Since about three-quarters of the greenhouse gas effect is due to carbon dioxide, the main mitigation priority should be to reduce emissions of carbon dioxide. Most carbon dioxide emissions come from burning fossil fuels, so the reduction of energy-related carbon dioxide emissions is the main priority on the mitigation agenda. If carbon dioxide levels could be held to below 450 ppm (parts per million), it would be likely (but not certain) to contain the rise in temperature to less than 2°C limit.

The decarbonisation of economic activity

The term decarbonisation refers to a large reduction of carbon dioxide per value of gross world product. Since most of the carbon dioxide comes from burning fossil fuels, a sharp reduction in the use of fossil fuels or a large-scale system to capture and sequester the carbon dioxide is needed. A decarbonised economy (also known as a low-carbon economy or low-fossil-fuel economy) is an economy based on power sources that either do not use carbon-based fuels or limits their use. Such economies – where economic growth has theoretically been 'decoupled' from fossil fuel use – would therefore have a minimal output of carbon dioxide emissions into the atmosphere.

There are three key steps being taken towards decarbonisation, each led by private and/or public sector players operating at different spatial scales:

1   Energy efficiency to achieve much greater output per unit of energy input. Much can be saved in heating, cooling and ventilation of buildings, and electricity use by appliances. Such measures might be adopted voluntarily by citizens for ethical reasons, or because of regulation by the state.
2   Reduce the emissions of carbon dioxide per megawatt hour of electricity generated. This involves increasing dramatically the amount of electricity generated by zero-emission energy, such as wind and solar power, while reducing the production of energy based on fossil fuels. It may also involve carbon capture and sequestration (see next section, page 173). The state may encourage these steps through the use of financial incentives or via punitive taxes on carbon use. Large transnational corporations (TNCs), such as BP and Shell, are important players in renewables research too – because these companies understand that their own longevity may depend upon a decarbonised business model.
3   Fuel shift, from direct use of fossil fuels to electricity based on clean primary energy sources. This kind of substitution of fossil fuels by clean energy can happen in many sectors. Internal combustion engines in cars could be replaced by electric motors. Battery-powered vehicles could be recharged on a renewable power grid. Large car manufacturing TNCs such as Volvo and Ford are key players developing these new technologies.

The shift to renewable energy sources

Table 6.1 compares renewable energy capacity at the beginning of 2004 to the end of 2016. For both years, hydro-electricity dominated renewable energy production, but most other sources of renewable energy have grown at a faster rate. Overall renewable power capacity almost doubled in the time period covered by Table 6.1. The newer sources of renewable energy making the largest contribution to global energy supply are wind power and biofuels.

Pearson Edexcel

AQA

OCR

KEY TERMS

Decarbonisation The reduction or removal of carbon dioxide from energy sources.

Decarbonised economy An economy based on low carbon power sources that therefore has a minimal output of greenhouse gas (GHG) emissions, specifically carbon dioxide.

	Unit	Start 2004	End 2016
New investment (annual) in renewable power and fuels	billion US$	39.5	241.6
Power			
Renewable power capacity (total, not including hydro)	gigawatts	85	5921
Renewable power capacity (total, including hydro)	gigawatts	800	2017
Hydropower capacity (total)	gigawatts	715	1096
Bio-power capacity	gigawatts	<36	112
Bio-power generation	terawatt hours	227	504
Geothermal power capacity	gigawatts	8.9	13.5
Solar PV capacity (total)	gigawatts	2.6	303
Concentrating solar thermal power (total)	gigawatts	0.4	4.8
Wind power capacity (total)	gigawatts	48	487
Heat			
Solar hot water capacity (total)	gigawatts	98	456
Transport			
Ethanol production (annual)	billion litres	28.5	98.6
Biodiesel production (annual)	billion litres	2.4	30.8

▲ **Table 6.1** Capacity of renewable energy sources, 2004 and 2016

Source: *Renewables 2017 Global Status Report*

The main drawback to the new alternative energy sources is that they invariably produce higher cost electricity than traditional sources. However, the cost gap with non-renewable energy is narrowing. As the analysis which follows shows, there are other drawbacks to consider too, either in terms of the way other alternative energy sources impact on other physical systems, or the extent to which they are truly carbon neutral.

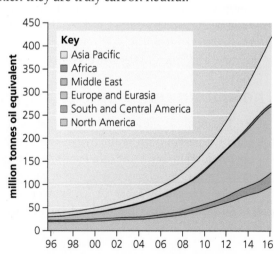

▶ **Figure 6.2** Renewable energy consumption by world region, 1996–2016. The newer sources of renewable energy making the largest contribution to global energy supply are wind power, solar and biofuels. Renewable energy in power generation (not including hydro-electric) grew by 14.1 per cent in 2016, slightly below the ten-year average, but the largest increase on record. Asia Pacific contributed 60 per cent of growth, with China overtaking the United States to become the world's largest renewable power producer

Hydro-electric power and water cycle issues

Hydro-electric power (HEP) is by far the most important source of renewable energy, use of which requires significant modifications to drainage basin water cycle storage and flow patterns (see page 89). However, most of the best HEP locations are already in use, so the scope for more large-scale development is limited. In many countries though, there is scope for small-scale HEP plants to supply local communities. However, global consumption of hydro-electricity increased from 687.5 million tonnes oil equivalent in 2006 to 910.3 million tonnes in 2016.

In 2016, the countries with the largest share of the world total were: China (28.9 per cent), Canada (9.7 per cent), Brazil (9.6 per cent) (Figure 6.3) and the USA (6.5 per cent).

Although HEP is generally seen as a clean form of energy, it is not without its problems:

- Large dams and power plants can have a huge negative visual impact on the environment.
- These dams and power plants may obstruct the river for aquatic life.
- There may be a deterioration in water quality.
- Large areas of land may need to be flooded to form the reservoir behind the dam.
- Submerging large forests rather than clearing the trees can release significant quantities of methane, a greenhouse gas.

In these and other ways, carbon cycle and water cycle management intersect with one another. Further water cycle issues associated with dam construction are explored on pages 89–91.

▲ **Figure 6.3** Inside the Itaipu hydro-electric power plant, Brazil

The environmental implications of wind power

Wind power is arguably the most important of the new renewable sources of energy (Figure 12.13). The worldwide capacity of wind energy is approaching 400,000 megawatts, a very significant production mark (Figure 6.5). Global wind energy is dominated by a relatively small number of countries. China is currently the world leader, with 31 per cent of global capacity, followed by the USA, Germany, Spain and India. Together, these five countries account for almost 72 per cent of the global total. In the last five years, for the first time ever, more new wind power capacity was installed in developing countries and emerging economies than in the developed world.

▲ **Figure 6.4** Wind farm in northern Spain

Country	Megawatts	Share (%)
PR China	114 763	31.1
USA	65 879	17.8
Germany	39 165	10.6
Spain	22 987	6.2
India	22 465	6.1
United Kingdom	12 440	3.4
Canada	9 694	2.6
France	9 285	2.5
Italy	8 663	2.3
Brazil	5 939	1.6
Rest of the world	58 275	15.8
Total Top 10	**311 279**	**84.2**
World total	**369 553**	**100**

Source: GWEC

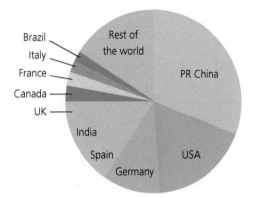

▲ **Figure 6.5** Global wind power capacity, end 2014

Costs of generating electricity from wind today are only about 10 per cent of what they were twenty years ago due mainly to advances in turbine technology. Wind energy operators argue that costs should fall further due to (a) further technological advances, and (b) increasing economies of scale. Table 6.2 summarises the advantages and disadvantages of wind power.

Advantages	Disadvantages
● A renewable source of energy that can produce reasonable levels of electricity with current technology.	● Growing concerns about the impact on landscapes as the number of turbines and wind farms increases.
● Advances in wind turbine technology over the last decade have reduced the cost per unit of energy considerably.	● NIMBY (not in my back yard) protests with people concerned about the impact of local turbines adversely affecting the value of their properties.
● Suitable locations with sufficient wind conditions can be found in most countries.	● The hum of turbines can be disturbing for both people and wildlife.
● Wind energy has reached the take-off stage both as a source of energy and as a manufacturing industry.	● Debate about the number of birds killed by turbine blades.
● Flexibility of location with offshore wind farms gaining in popularity.	● TV reception can be affected by wind farms.
● Repowering can increase the capacity of existing wind farms.	● The development of wind energy has required significant government subsidies – some people argue that this money could have been better spent elsewhere (opportunity cost).
● Significant public support for a renewable source of power, although this may be waning to an extent.	● Many wind farms are sited in coastal locations where land is often very expensive.

▲ **Table 6.2** The advantages and disadvantages of wind power

 KEY TERM

Repowering The process of replacing older power stations with newer ones that either have a greater full-load sustained output or are more efficient, resulting in an overall increase in power generated.

Renewable sources of energy such as wind are often promoted as being 'carbon neutral', but to what extent is this true?

- Although no carbon is released during the operation of wind turbines, carbon dioxide will be released during their production. Wind turbines are mainly made of steel, with concrete bases. Steel is made with coal, to supply the carbon in the alloy and also to provide heat for smelting the iron ore.
- Concrete is also often made using coal. It is calculated that approximately half a tonne of coal is needed to make a tonne of steel, 25 tonnes to make the concrete base: 2-megawatt wind turbine blades weigh around 250 tonnes, and so overall around 150 tonnes of coal are needed to produce each wind turbine. Building 350,000 wind turbines, to keep up with increasing energy demand, would require 500 million tonnes of coal a year (equivalent to half the EU's coal-mining output), significantly adding to carbon dioxide emissions.
- It would appear, therefore, that wind turbines are far from carbon neutral, and that using natural gas (which has lower emissions of carbon dioxide than other fossil fuels – see Chapter 5, page 134) may be a more effective way of reducing carbon emissions than wind technology.

Biofuels, biodiversity and the carbon cycle

Biofuels are fossil fuel substitutes that can be made from a range of agri-crop materials, including oilseeds, wheat, corn and sugar. They can be blended with petrol and diesel.

The main methods of producing biofuels are:

- crops that are high in sugar (sugar cane, sugar beet, sweet sorghum) or starch (corn/maize) are grown and then yeast fermentation is used to produce ethanol
- plants containing high amounts of vegetable oil (such as oil palm, soybean and jatropha) are grown, and the oils derived from them are heated to reduce their viscosity; they can then be burned directly in a diesel engine, or chemically processed to produce fuels such as biodiesel
- wood can be converted into biofuels such as woodgas, methanol or ethanol fuel
- cellulosic ethanol can be produced from non-edible plant parts, but costs are not economical at present – this method is seen as the potential second generation of biofuels.

Ethanol is the most common biofuel globally (over 90 per cent of total biofuel production), particularly in the USA and Brazil, which produce 87 per cent of the world total (Figure 6.6). Most existing petrol engines can run on blends of up to 15 per cent ethanol. In the USA, about 40 per cent of the maize crop is used to produce ethanol. However, production in the European Union and China is growing significantly.

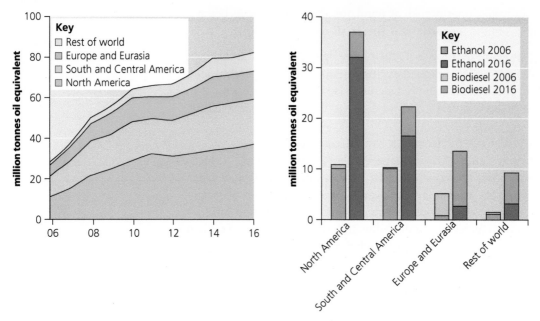

▲ **Figure 6.6** Global biofuel production, 2006–16. Global biofuels production rose by 2.6 per cent in 2016, well below the ten-year average of 14.1 per cent, but faster than in 2015 (0.4 per cent). The USA provided the largest increment. Global ethanol production increased by only 0.7 per cent, partly due to falling production in Brazil. Biodiesel production rose by 6.5 per cent with Indonesia providing more than half of the increment

Increasing investment is taking place in research and development of the so-called 'second generation' biodiesel projects, including algae and cellulosic diesel. Other important trends in the industry are a transition to larger plants and consolidation among smaller producers.

To what extent is biofuel successful in decarbonising the economy? Although biofuel can be considered carbon neutral (see page 160), there are implications for carbon storage of removing natural biomes and replacing them with biofuels.

- In terms of carbon storage, the productivity of biofuel crops will be much less than natural forest ecosystems and so carbon storage will be much reduced.
- Another problem with biofuel is that, in many cases, the production of the biomass for fuel competes with food production. Approximately 100 million tonnes of grain are used for biofuels. As more grain is used for biofuel, less grain (and land) is used for the production of food for human use.
- Conversion of forest or grassland to crop production also has a significant effect on biodiversity due to habitat loss. Many current biofuel crops are well suited for tropical areas, creating an economic incentive to convert biodiverse natural ecosystems into monoculture biofuel plantations.

The environmental implications of geothermal electricity

Geothermal energy is the natural heat found in the Earth's crust in the form of steam, hot water and hot rock. Rainwater may percolate several kilometres below the surface in permeable rocks, where it is heated due to the Earth's geothermal gradient: the rate at which temperature rises as depth below the surface increases. The average rise in temperature is about 30°C per kilometre, but the gradient can reach 80°C near plate boundaries.

This source of energy can be used to produce electricity, or its hot water can be used directly for industry, agriculture, bathing and cleansing (Figure 6.7). For example, in Iceland hot springs supply water at 86°C to 95 per cent of the buildings in and around Reykjavik. At present, virtually all the geothermal power plants in the world operate on steam resources, and they have an extremely low environmental impact.

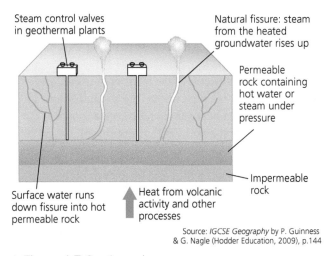

Source: *IGCSE Geography* by P. Guinness
& G. Nagle (Hodder Education, 2009), p.144

▲ **Figure 6.7** Geothermal power

First begun in Larderello, Italy, in 1904, total world installed geothermal capacity passed 12,000 megawatts by the end of 2013. This is enough electricity to meet the needs of over 70 million people. The USA is the world leader in geothermal electricity, with plants in Alaska, California, Hawaii, Nevada and Utah. Total production accounts for about 0.4 per cent of the electricity used in the USA. Other leading geothermal-electricity-using countries are the Philippines, Indonesia, Mexico, Italy, New Zealand, Iceland and Japan.

Geothermal power has low environmental impact and plants occupy relatively small areas of land. Geothermal plants reliably produce power and generation is not dependent on weather conditions (unlike wind and solar power). There are some limitations, however: there are few locations worldwide where significant amounts of energy can be generated, and total generation remains small. Some of the best locations are far from where the energy could be used and installation costs of plant and piping are relatively high. In terms of reducing carbon emissions, however, geothermal provides an extremely effective option: no combustion is involved so geothermal plants will produce nearly zero carbon emissions, helping to offset energy-related carbon dioxide from other sources.

▲ **Figure 6.8** Solar electricity generated by photovoltaic panels in Spain

Solar power and solar panel manufacturing issues

From a relatively small base, the installed capacity of solar electricity is growing rapidly. Experts say that solar power has huge potential for technological improvement, which could make it a major source of global electricity in years to come (Figure 6.8). In 2000, global solar capacity was only 1275 megawatts. It grew to 5085 megawatts in 2005 and 40,183 in 2010. Global solar power capacity passed the milestone of 100,000 megawatts in 2012.

Solar electricity is currently produced in two ways:

- Photovoltaic (PV) systems – solar panels that convert sunlight directly into electricity.
- Concentrating solar power (CSP) systems – use mirrors or lenses and tracking systems to focus a large area of sunlight into a small beam. This concentrated light is then used as a heat source for a conventional thermal power plant. The most developed CSP systems are the solar trough, parabolic dish and solar power tower.

Each method varies in the way it tracks the Sun and focuses light. In each system, a fluid is heated by the concentrated sunlight, and is then used for power generation or energy storage.

Another idea being considered is to build solar towers. Here, a large glassed-in area would be constructed with a very tall tower in the middle. The hot air in this 'greenhouse' would rise rapidly up the tower, driving turbines along the way.

Traditional solar panels comprise arrays of photovoltaic cells made from silicon. These cells absorb photons in light and transfer their energy to electrons, which form an electrical circuit. However, standard solar panels:

- are costly to install
- have to be tilted and carefully positioned so as not to shade neighbouring panels.

The silicon required for PV modules is extracted from quartz sand at high temperatures, which is the most energy intensive phase of PV module production, accounting for 60 per cent of the total energy requirement. CO_2 emissions for photovoltaic power systems are currently 58 gCO_2e/kWh (grams of CO_2 equivalent per kilowatt-hour of electricity generated – i.e. the amount of carbon dioxide, or other GHGs, emitted generating one kilowatt of electricity per hour).

A number of companies are now developing a new technique to manufacture solar panels, such as Canadian Solar Inc. (a company based in Toronto) and First Solar Inc. (an Arizona-based manufacturer of solar panels with global operations). These operations involve using different materials and building them in very thin layers or films, almost like printing on paper, to produce the photovoltaic effect. The cost of production is reduced because the layers or films use less material, and they can be deposited on bases such as plastic, glass or metal.

As well as a lower cost, this latest generation of solar panels has a lower carbon footprint, because less silicon is used. Chinese TNCs are leading developers of these new technologies. By 2010, four of the top five solar cell producers were based there. For example, Hanergy Thin Film Power Group Ltd is a Hong-Kong-based manufacturer which produces flexible and lightweight thin-film PV panels. China, although the number one carbon emitter, may also be key player in reducing carbon flows eventually through solar research and development.

Finally, there are intersecting economic and political issues for geographers to consider when exploring the use of new solar panels:

- EU state and US governments have been critical of the way the Chinese government currently subsidises solar panel production before 'dumping' the products on European and US markets; they say that China is breaking World Trade Organisation (WTO) rules that global economic systems depend on.
- In 2018, the Trump administration imposed import tariffs on Chinese panels in an attempt to protect US solar power industries.
- As a result, future use of solar panels will be determined not only by our urgent need to decarbonise energy but also by the operation of global trade systems.

Tidal power and its carbon footprint

Although currently in its infancy, a study by the Electric Power Research Institute has estimated that as much as 10 per cent of US electricity could eventually be supplied by tidal energy. This potential could be equalled in the UK and surpassed in Canada.

Tidal power plants act like underwater windmills, transforming sea currents into electrical current. Tidal power is more predictable than solar or wind power, and the infrastructure is less obtrusive, but start-up costs are high. The 240-megawatt Rance facility in north-western France is the only utility-scale tidal power system in the world. However, the greatest potential is Canada's Bay of Fundy in Nova Scotia. A pilot plant was opened at Annapolis Royal in 1984, which at peak output can generate 20 megawatts. More ambitious projects at other sites along the Bay of Fundy are under consideration, but there are environmental concerns. The main concerns are potential effects on fish populations, levels of sedimentation building up behind facilities and the possible impact on tides along the coast. In addition, tidal plants require large amounts of concrete and steel, which increase their carbon footprint (see also page 165) – most CO_2 is emitted during manufacture of the structural materials, and a wave converter device presently requires 665 tonnes of steel. Emissions for this type of marine technology is estimated between 25–50 gCO_2eq/kWh, roughly equivalent to CO_2 emissions from current PV systems.

Fuelwood use in developing countries

In developing countries and rural regions of many emerging economies, about 2.5 billion people rely on wood, charcoal and animal dung for cooking (Figure 6.9). Wood and charcoal are collectively called fuelwood, which accounts for just over half of global wood production. Fuelwood provides much of the energy needs for sub-Saharan Africa. It is also the most important use of wood in Asia.

According to the World Energy Outlook, 1.3 billion people were still living without access to electricity in 2012. This is equal to 18 per cent of the world's population. Nearly 97 per cent of those without access to electricity live in sub-Saharan Africa or in Asia. The largest populations without electricity are in India, Nigeria, Ethiopia, Bangladesh, Democratic Republic of Congo and Indonesia.

In developing countries, the concept of the energy ladder is important. Here, a transition from fuelwood and animal dung to 'higher-level' sources of energy occurs as part of the process of economic development (and so this could reduce carbon footprints if the higher level source is HEP). Forest depletion is therefore initially heavy near urban areas but slows down as cities become wealthier and change to other forms of energy. It is the more isolated rural areas that are most likely to lack connection to an electricity grid. It is in such areas that the reliance on fuelwood is greatest. Wood is likely to remain the main source of fuel for the global poor in the foreseeable future. The collection of fuelwood does not cause deforestation on the same scale as the clearance of land for agriculture, but it can seriously deplete wooded areas, reducing the carbon store and increasing carbon dioxide emissions.

▲ **Figure 6.9** Animal dung being dried for fuel in India

Carbon capture and storage

▶ *What techniques can be used to increase carbon stores and rebalance carbon fluxes so that there are net influxes and reduced effluxes into the atmosphere?*

Removing carbon dioxide from the atmosphere (carbon capture) and then storing it so that it is no longer part of the global carbon cycle (carbon storage) is another important mitigation approach.

Afforestation

Afforestation involves planting trees in deforested areas or places that have never been forested. New trees act as carbon sinks and can therefore help with climate change mitigation:

 KEY TERM

Afforestation The establishment of forests in an area where there was no previous tree cover.

- The UN-REDD Programme, launched in 2008, is the United Nations Initiative on Reducing Emissions from Deforestation and Forest Degradation (REDD) in low-income countries. REDD provides incentives for developing countries to conserve their rainforests by placing a monetary value on forest conservation. This is an important example of successful global governance. REDD stresses the role of conservation, the sustainable management of forests and the increase of forest carbon stocks. By June 2014, total funding had reached almost US$200 million. Norway is the leading donor country.
- The UK Forestry Commission was established in 1919, following the end of the First World War, to increase timber supplies through a policy of land-use changes, a rare early example of human action bringing a *positive* change in carbon storage. Marginal areas of grassland, heather and moorland were used to grow coniferous forest, for example in the Brecon Beacons (Wales) and Isle of Arran (Scotland).
- New monoculture of commercial trees, such as coniferous plantations in the UK, can increase carbon storage if it replaces grassland. However, it may store less carbon than natural forest biome communities do. In addition, monoculture forest lacks biodiversity and provides few habitats for other plant and animal species to occupy.
- Individual citizens can play an active role in afforestation through the practice of carbon offsetting, which is another widely used mitigation strategy that aims to marry business principles with environmental goals. Our everyday actions, such as driving, flying and heating buildings, consume energy and produce carbon emissions. Carbon offsetting is a way of compensating for your emissions, although the amount of carbon removed from the atmosphere is only really a drop in the ocean compared to the amount of emissions being produced. For example, offsetting the UK's annual GHG emissions would require planting an area of forest the size of Devon and Cornwall combined every year and maintaining these

forests forever. In addition, many argue that carbon stored in trees or biological carbon is not equivalent to fossil carbon, as it will be released back into the atmosphere through fire, natural decay or harvesting.

ANALYSIS AND INTERPRETATION

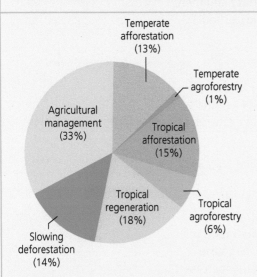

▲ **Figure 6.10** The potential of various land-management activities to mitigate global emissions of CO_2. Estimates provided by the IPCC suggest that a maximum mitigation of 100 PgC could be achieved between 2000 and 2050 (reproduced from The Royal Society, 2001)

Figure 6.10 shows different land management techniques to reduce global emissions of carbon dioxide.

(a) Outline two contrasting methods of reducing carbon emissions indicated in Figure 6.10.

GUIDANCE

This is a simple question that asks you to interpret data presented in the pie chart. It asks you to look for contrasting methods – think about contrasting flows of carbon and how different processes presented in the chart may impact them.

(b) Explain how tropical afforestation and slowing deforestation can be used to mitigate global emissions of CO_2.

GUIDANCE

This question asks you to apply your knowledge of photosynthesis and carbon capture and storage. Think about how these processes decrease atmospheric carbon dioxide stores and increase land-based stores.

(c) Suggest how agricultural management can lead to reduced atmospheric carbon dioxide concentrations.

GUIDANCE

Agroforestry combines agriculture with forestry, allowing the farmer to continue cropping while using trees for fodder, fuel and building timber: think about how these factors would impact flows and storages within the carbon cycle. In addition, think about how soil protection would increase carbon retention. What effect would increased use of fertilisers have in a system with little or no fertiliser and how would this improve carbon capture and storage?

Carbon capture and storage (or sequestration) (CSS)

Carbon capture and storage (CSS) is a technological innovation that allows carbon dioxide to be captured and stored instead of allowing it to accumulate in the atmosphere (see page 159, and Figure 6.11).

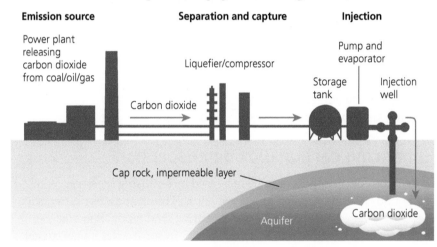

▲ **Figure 6.11** How carbon capture and storage (CCS) works

There are many technical and policy issues about the feasibility and cost-effectiveness of large-scale CCS technologies. Firstly, how costly will it be to capture carbon dioxide on a large scale? How costly will it be to ship the carbon dioxide by a new pipeline network and then store the carbon dioxide in some safe, underground geologic deposit? If the carbon dioxide is put underground, how certain are we that the carbon dioxide will stay where it is put, rather than returning to the surface and then into the atmosphere? Tens of billions of tons of carbon dioxide would have to be captured and stored each year for CCS to play the leading role in addressing carbon dioxide emissions. Is there enough room for all this carbon? There is relatively little research and development underway to test the economic and geologic potential for large-scale CCS. Table 6.3 shows an evaluation of CCS technology.

Strategy	Analysis	Evaluation
Carbon capture and storage (CCS)	CCS involves capturing carbon dioxide released by the burning of fossil fuels and burying it deep underground (Figure 6.11). This technique promises to be extremely important (given that coal will remain a very significant part of the global energy budget for years to come due to its abundance and low cost). CCS works in three stages. 1 The carbon dioxide is separated from power station emissions. 2 The gas is compressed and transported by pipeline to storage areas. 3 It is injected into porous rocks deep underground or below the ocean for permanent storage (geo-sequestration). CCS could make an enormous difference to the size of the anthropogenic carbon store. The IPCC estimates that CCS (1) has the potential to reduce coal power station emissions by up to 90 per cent, and (2) could provide up to half of the world's total carbon mitigation until 2100.	● So far the technology has been piloted at only a handful of coal-fired power stations worldwide: it is far from being a mature technology. ● CCS will be expensive because the technology is complex and still being developed. ● There is uncertainty over how successful it will be. For carbon dioxide to remain trapped underground there must be no possibility of any leak to the surface. The gas cannot be allowed to re-enter the atmosphere once it has been removed. ● Pilot projects in the UK were cancelled recently due to rising costs (of over US$1 billion). The plan had been for carbon to be transported by a pipeline to the North Sea and stored in depleted gas reservoirs. The UK government has cut public spending in many areas because of the global financial crisis and its aftereffects.

▲ **Table 6.3** An evaluation of carbon capture and storage technology

Source: Simon Oakes, (2017), *Geography Study and Revision Guide: Global Change*, Hodder

Enhancing carbon dioxide absorption

Carbon dioxide absorption can be increased by fertilising the ocean with compounds of iron, nitrogen and phosphorus. This introduces nutrients to the upper layer of the oceans, increases marine food production and removes carbon dioxide from the atmosphere. In some cases it may trigger an algal bloom. The algae trap carbon dioxide and sink to the ocean floor.

Sperm whales transport iron from the deep ocean to the surface during prey consumption and defecation. Increasing the number of sperm whales in the Southern Ocean could help remove carbon from the atmosphere.

In some locations, upwelling currents bring nutrients to the surface (e.g. off the coast of Peru). These support large-scale fisheries and also help to lock up carbon in plankton and algal blooms. Artificial upwelling can be produced by devices that help pump water to the surface. Ocean wind turbines can also cause upwellings. These can then support plankton blooms which help lock up carbon. However, these are costly to build and run.

Climate change adaption and resilience

▶ *How can countries adapt to a warmer world, and what factors affect the ability of an environment or society to adapt to climate change?*

At the Paris Climate Change Conference, COP21 (page 157), the global community agreed to provide continued and enhanced international support to developing countries to help them adapt to the impacts of climate change. Some strengths and weaknesses of the outcomes from COP21 and the Paris Agreement are shown in Table 6.4.

Strengths	Weaknesses
● Getting 195 countries to agree on anything is a major achievement! ● From a scientific perspective, COP21 gives us hope that so-called 'dangerous climate change' across all continents can be avoided. ● From an economic perspective, the agreement gives countries time to decarbonise their economies 'without sacrificing economic prosperity on the altar of environmental wellbeing'. ● From a political perspective, it allows all governments to hold each other to frequent account regarding emissions levels and targets. Ten-year reviews, for instance, would be too infrequent. ● Finally, from the perspective of poorer, low-lying countries like Bangladesh, elements of COP21 promise a degree of justice for those adversely affected by wealthier countries' previous GHG emissions.	● Massive GHG reductions will be required by 2050 to keep the temperature increase below 2°C; some countries may find it too expensive to phase out things like coal-fired power stations over the required timescale. ● If a country is hit by an economic recession or experiences political change, its priorities may alter and climate could slip down the domestic agenda (compare President Trump's commitment to that of President Obama, for instance). ● Replacing fossil fuel economics with renewable energy ones sufficient to maintain decent lifestyles requires technologies that have yet to be invented! ● The Paris Agreement is 'merely a statement of intent, albeit an important one. The key to its success or failure lies in... the fine details of how countries respond when one or more fail to honour their commitments.'

▲ **Table 6.4** Strengths and weaknesses of the COP21 'Paris Agreement'

Source: Noel Castree, *Geography Review* 30(1)

Adaptation strategies reduce the impact of global warming, and allow communities to live with the consequences of climate change.

Adaptation strategies include:

● building flood defences, such as levees or dikes
● using desalinisation plants to replace freshwater losses
● planting of crops in previously unsuitable areas

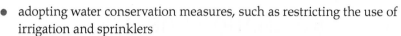

- adopting water conservation measures, such as restricting the use of irrigation and sprinklers
- exploiting areas that have become more productive for crops through climate change
- developing crops better adapted to areas impacted by climate change, such as those that are drought-resistant, either through selective breeding or genetic modification
- using green roof systems that cool the building through transpiration and reflection of incident sunlight radiation
- vaccination programmes.

Some strategies can be acknowledged as both adaptive and mitigating, for example a 'green roof' both reduces the impact of climate change by cooling the building (adaptation) and reduces cause of climate change by reducing carbon emission (mitigation).

It is possible to reduce human emissions of greenhouse gases substantially. The technologies are within reach, and measures such as energy efficiency, low-carbon electricity and fuel switching (e.g. electrification of buildings and vehicles) are all possible. Nevertheless, even with these, carbon dioxide will continue to rise for a number of decades. By the time that the oceans warm, they are likely to add a further 0.6°C to global temperatures. Thus, as well as trying to mitigate climate change, humanity needs to adapt to climate change as well. Adaptation strategies can be used to reduce adverse effects and maximise any positive effects.

Adaptive capacity varies from place to place of course because of how it depends on financial and technological resources. Developed countries can provide economic and technological support to emerging economies and developing countries. In agriculture, crop varieties must be made more resilient to higher temperatures and more frequent floods and droughts. Cities need to be protected against rising ocean levels and greater likelihood of storm surges and flooding. The geographic range of some diseases, such as malaria, will spread as temperatures rise. More widespread vaccination programmes will be needed to deal with the spread of such diseases. To cope with changes in the supply of (and demand for) water, more desalinisation plants will be required. These are expensive and some developing countries may struggle to meet the demand for fresh water.

Implementing mitigation and adaptation strategies at a global scale

The ability to implement mitigation and adaptation strategies may vary from one country to another. Some countries may not have the political

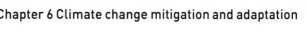

will to reduce the use of fossil fuels, in part due to their population's unwillingness to change their lifestyle or restrict economic growth. Political corruption may, in some instances, cause governments to redirect funds from a sustainability agenda to projects that enrich individuals at the expense of the state.

Developing countries may lack the financial support to allow them to fund new technologies and the infrastructure needed to adopt a more sustainable lifestyle. Some countries may depend upon others for knowledge transfer or technological assistance to implement resolutions.

Religious, cultural or political norms may limit some countries and hold them back from making the necessary changes to reduce their carbon footprint. Inadequate or limited education in some countries may promote or limit the perception of environmental threats and approaches to management within the population.

The geographical location of some countries may place them at greater or more immediate risk from the impacts of climate change, such as low-lying islands and tropical storm-prone nations. Conversely, the location of a country may provide it with greater opportunities for mitigation, such as available sources of alternative energy (e.g. wind power; tidal power).

Some countries may perceive greater immediate priorities over and above the issues posed by climate change, such as war in Syria and poverty in Somalia.

National adaptation programmes

The main content of national adaptation programmes of action (NAPAs) is a list of ranked priority adaptation activities and projects. It provides a process for Least Developed Countries (LDCs) to identify priority activities that respond to their urgent and immediate needs to adapt to climate change. NAPAs focus on needs for which further delay could increase vulnerability or lead to increased costs at a later stage. NAPAs use existing information, are action-orientated and country-driven. The steps for the preparation of the NAPAs include:

- synthesis of available information
- assessment of vulnerability to current climate and extreme events
- identification of key adaptation measures as well as criteria for prioritising a selection of activities.

By 2008, the UNFCCC had received NAPAs from 39 developing countries.

CONTEMPORARY CASE STUDY: TACKLING CARBON EMISSIONS: NON-GOVERNMENTAL AND CORPORATE ACTION IN THE USA

Many high-profile US politicians, including President Trump, are climate change sceptics. As a result, citizens who care about the issue believe it is more important than ever to act in whatever ways they can to reduce carbon emissions (no matter how small the scale of action). Table 6.5 shows a selection of non-governmental US players and analyses the actions they have taken in response to climate change.

This raises the question of who has the greatest power to make a difference in relation to climate change in the USA and other countries. Is it the United Nations? The US White House? TNCs? President Trump's withdrawal of the USA from the Paris accord in 2017 may ultimately have limited impact, as many argue that the economic incentives for TNCs to move towards renewable sources of energy, and away from fossil fuels, will ultimately lead the USA towards a more sustainable future.

Players	Actions analysis
NextGen Climate (a civil society organisation)	• NextGen Climate in an environmental pressure group whose mission is to engage politically with millennials about the connected issues of climate issue and clean energy. They advise potential voters on which politicians share their environmental concerns. • The organisation has a field operation in Las Vegas which is viewed as being on the 'front line' of climate change in the USA: Las Vegas has doubled its consumption of water twice since 1985 and has suffered from severe droughts in recent years, including 2016.
Citizens	• In recent years, increasing numbers of Las Vegas citizens and garden businesses have begun to adapt to what they perceive to be a permanent change towards even more arid conditions. Homeowners favour drought-tolerant 'desert landscaping' and are abandoning water-hungry grass lawns. • In colder states, such as Montana, individual actions could involve turning down a home's thermostat by 1°C; this brings a 3 per cent reduction in total household energy use. This action cannot be forced through legislation and relies on action by individual citizens. Government can, however, play a role by educating people about the issues.
ExxonMobil (an oil and gas TNC)	• ExxonMobil has a long record of investing in CCS research. Despite having much to lose if people were to abandon the use of fossil fuels, some US energy companies, such as ExxonMobil, play an active role in supporting geoengineering. Indeed, a corporate strategy of supporting CCS technology may be essential to their long-term profitability: it would mean consumers can keep using oil and gas while trusting CCS to remove anthropogenic carbon from the atmosphere.
Seattle	• The city of Seattle has its own Climate Action Plan (CAP) which aims to make the city carbon neutral by 2050. Adopted in 2013, Seattle CAP focuses on city actions that reduce greenhouse emissions and also 'support vibrant neighbourhoods, economic prosperity, and social equity'. • Actions are focused on areas of greatest need and impact: road transportation, building energy and waste.

▲ **Table 6.5** Examples of civil society campaigning in the USA

Source: Simon Oakes, (2017), *Geography Study and Revision Guide: Global Change*, Hodder

Variations in levels of climate change risk

There is no doubt that climate change will have a serious impact on many places and communities. Levels of climate change risk and vulnerability will vary according to a person's location, wealth, social differences (age, gender, education) and risk-perception. As a result, resilience to climate change will vary according to levels of development and wealth, as well as location. Vulnerability to global climate change refers to the degree to which people are susceptible to, or unable to cope with, the adverse impacts of climate change. There are three main factors associated with vulnerability:

- Exposure – the degree to which people are exposed to climate change.
- Sensitivity – the degree to which they could be harmed by exposure to climate change.
- Adaptive capacity – the degree to which they could mitigate the potential harm by taking action to reduce their exposure or sensitivity.

Some locations are more at risk than others. These include low-lying islands, river mouths and valleys, coastal areas and regions that derive their water supplies from mountain glaciers. Many islands in the Indian Ocean and the Pacific Ocean are among the areas most vulnerable to climate change risks. These include the Kiribati, Tuvalu and the Marshall Islands, the Maldives, and Antigua and Nevis in the Caribbean. Much of the infrastructure and socio-economic activities of these islands is located along the coastline. An 80 cm rise in sea level could inundate about 66 per cent of Kiribati and the Marshall Islands. A 90 cm rise in sea level could inundate 85 per cent of Malé, capital of the Maldives. The problems these low-lying islands face include increased coastal erosion, saline intrusion into groundwater supplies, deterioration of coral reefs, out-migration of people and loss of income as a result of a decline in economic activities and infrastructure.

> **KEY TERM**
>
> **Resilience** The ability of an environment or society to adapt to changes that have a negative impact upon them, such as climate change.

 # Evaluating the issue

▶ *To what extent is a warmer world inevitable?*

Identifying possible contexts, data sources and criteria for the evaluation

The focus of this chapter's debate is the extent to which a warmer world is now inevitable. In order to answer this question, scientists must take into consideration many different indicators and uncertainties relating to anthropogenic climate change; while additionally factoring in the extent to which human actions could influence the trajectory towards a warmer world, either positively or negatively.

Increasing levels of carbon dioxide in the atmosphere (Table 5.6, page 150) – and the established correlation between increased GHG emissions and the enhanced greenhouse effect – certainly suggest that a warmer world is inevitable.

Pearson Edexcel   AQA   OCR   WJEC/Eduqas

However, two important uncertainties surround this assertion:

- The phrase 'a warmer world' carries a wide range of meanings; because of the complexities involved in climate modelling and systems analysis (see pages 14–15), IPCC scientists have generated a range of different predictive models, features of which are subject to their own inherent uncertainties in terms of *exactly how much warming will occur*.
- The word 'inevitable' is ambiguous *with respect to the temporal scale which it invokes*. Chapter 1 explored Earth system modifications occurring over timescales ranging from days to hundreds of millions of years. We know that the climate inevitably warms and cools over time because of naturally occurring glacial and interglacial cycles. However, IPCC scientists project an unprecedentedly rapid rise in temperature during the century; a range of evidence suggests that process is already well underway.

Evaluating the view that a warmer world *is* inevitable

Taken together, a range of evidence suggests that Earth's climate is currently warming and changing; the US National Oceanic and Atmospheric Administration (NOAA) says the signs are 'unmistakable'.

- In 2015, global mean surface temperature (GMST) reached a new record high of +0.87°C relative to the 1951–80 average GMST.
- The ten warmest years since 1880 have all been since 1998. The prevailing viewpoint held by the large majority of the world's climate scientists is that global warming is caused by increased greenhouse gas (GHG) emissions, and that humans are the cause of this. The UK Meteorological Office says that the signs of warming 'have human fingerprints' all over them. In the fifth climate change assessment (2013), Intergovernmental Panel on Climate

Change (IPCC) scientists reported they were 'virtually certain' that humans are to blame for 'unequivocal' global warming.

The key steps in the argument are as follows:

- There has been a marked increase in atmospheric carbon storage. Carbon dioxide emissions have been rising since 1750, the start of Europe's industrial revolution, from a level of 280 ppm (parts per million) to 406 ppm at the start of 2017. This represents an increase of around 45 per cent. Even more worryingly, if you convert other GHGs (methane, nitrous oxide) into their equivalent amounts of CO_2, then you find that we have reached a level in excess of 470 ppm of CO_2 equivalents.
- This is because population growth and economic development have led to: worldwide use of carbon-rich fossil fuels as an energy source; widespread deforestation; cement manufacturing; enhanced methane emissions derived from livestock (bovine flatulence) and the decomposition of organic wastes in landfill sites.

Much of the evidence for the greenhouse effect has been taken from ice cores dating back 160,000 years. These show that the Earth's temperature closely paralleled the levels of CO_2 and methane in the atmosphere (Figure 6.12). Calculations indicate that changes in these greenhouse gases were part, but not all, of the reason for the large (5–7°) global temperature swings between ice ages and interglacial periods.

Accurate measurements of the levels of CO_2 in the atmosphere began in 1957 in Hawaii. The site chosen was far away from major sources of industrial pollution and shows a good representation of unpolluted atmosphere. Measurements have been taken offshore at Mauna Loa every year since the 1950s and these show a rise each year of around 1–2 ppm.

Older evidence comes from 'fossil air' trapped in ice. Permanent ice in high mountains and polar ice caps has built up from snow falling over hundreds of thousands of years. Cores between 3–4 kilometres long have been extracted from the Vostok ice sheet in Antarctica, producing ice that is more than half a million years old and containing bubbles of ancient air. After analysing the evidence, scientists believe that carbon dioxide and methane concentrations are now higher than at any time in the last 800,000 years (Figure 6.12).

Studies show that the level of CO_2 between 10,000 years ago and the mid-nineteenth century was stable, at about 270 ppm. By 1957, the concentration of CO_2 in the atmosphere was 315 ppm, and it has since risen to about 360 ppm. Most of the extra CO_2 has come from the burning of fossil fuels, especially coal, although some of the increase may be due to the disruption of the rainforests. For every tonne of carbon burned, 4 tonnes of CO_2 are released. By the early 1980s, 5 gigatonnes (5000 million tonnes, or 5 Gt) of fuel were burned every year. Roughly half the CO_2 produced is absorbed by natural sinks, such as vegetation and plankton.

A so-called 'hockey stick' trend line can be seen in carbon dioxide data: there was gradual growth up to the early 1900s and then very steep growth (Figure 6.12).

Evaluating the view that a warmer world is *not* inevitable

Although the evidence for a warming world, due to anthropogenic activities, seems irrefutable, the models used to predict how global surface temperatures will change during the twenty-first century contain many uncertainties. Models, by their very nature, are simplifications of reality (see page 14), whereas climate change science is far from simple. It is a very complex issue for a number of reasons:

- Scale – it includes the atmosphere, oceans and land masses across the world.
- Interactions between these three areas are complex.
- It includes natural as well as anthropogenic forces.
- There is uncertainty over the operation of feedback loops.
- Many of the processes are long term and so the impact of changes may not yet have occurred.

Climate change presents us with a problem which – because of its complexity – defies attempts to establish clearly what its exact extent and effects are likely to be. Unfortunately, this reasonable uncertainty is seized upon by climate change sceptics as a reason to avoid taking any action to reduce carbon dioxide and other GHG emissions.

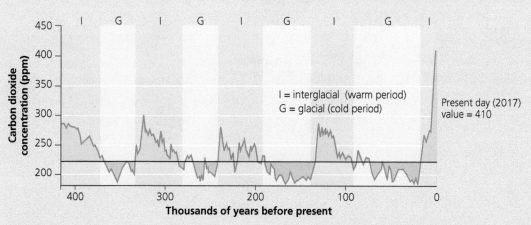

▲ **Figure 6.12** Atmospheric carbon dioxide concentrations measured from Vostok ice cores and recent Mauna Loa data

Figure 6.13 provides a useful summary of some of the many different influences on climate change processes and the high levels of complexity and uncertainty which accompany any attempt to model the implications of rising carbon emissions for life on Earth.

The scientific consensus is that anthropogenic warming is occurring, although there is still some political and sociological dissent. Some people claim that global warming is due to normal variations in the Earth's climate, for example. All models, however, predict warming in varying degrees: any uncertainly is about the magnitude of warming rather than about warming *per se*. Figure 1.15 (page 15) shows two different climate change pathways modelled by the Intergovernmental Panel on Climate Change (IPCC). In each case a range of temperature rises are shown, with differing implications for life on Earth. The GMST rise prediction for a high-emissions scenario varies from 3°C to 5°C. Among many other things, this reflects uncertainty around the strength and timing of possible positive feedback loops linked with Arctic permafrost thawing.

As well as uncertainties about the future trajectory of the enhanced greenhouse effect, changes in human behaviour may lead to a reduction in carbon emissions and the inevitability of a significantly warmer world may become less certain. Changes in eating habits provide one realistic opportunity for

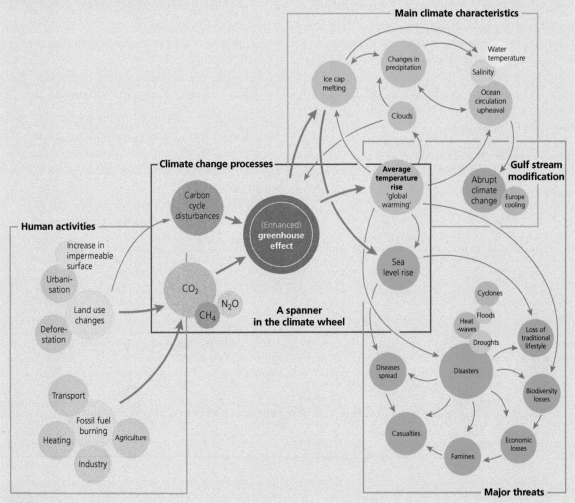

▲ **Figure 6.13** Understanding the complexity of global systems and feedback loops affecting climate change

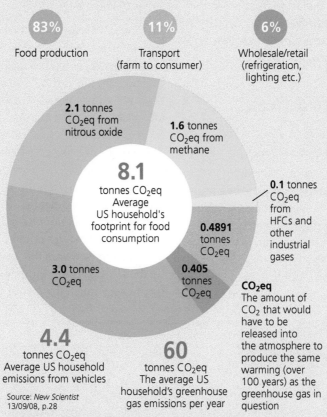

Household greenhouse gas emissions from food account for almost twice those produced by vehicle use. Most of this comes from the food production process itself, rather than food miles, as is often believed

83%
Food production

11%
Transport
(farm to consumer)

6%
Wholesale/retail
(refrigeration,
lighting etc.)

2.1 tonnes
CO₂eq from
nitrous oxide

1.6 tonnes
CO₂eq from
methane

8.1
tonnes CO₂eq
Average
US household's
footprint for food
consumption

0.1 tonnes
CO₂eq
from
HFCs and
other
industrial
gases

0.4891
tonnes
CO₂eq

0.405
tonnes
CO₂eq

3.0 tonnes
CO₂eq

CO₂eq
The amount of
CO₂ that would
have to be
released into
the atmosphere to
produce the same
warming (over
100 years) as the
greenhouse gas in
question

4.4
tonnes CO₂eq
Average US household
emissions from vehicles

Source: *New Scientist*
13/09/08, p.28

60
tonnes CO₂eq
The average US
household's greenhouse
gas emissions per year

▲ **Figure 6.14** Comparison of household greenhouse gas emissions from food and vehicle use

humanity to reduce anthropogenic planetary warming. It has been estimated that food production and consumption accounts for up to twice as many greenhouse emissions as driving vehicles. Figure 6.14 shows US data published in the *New Scientist*. The average US household's footprint for food consumption is 8.1 tonnes of carbon dioxide equivalent, compared with 4.4 tonnes from driving.

In 2000, annual global meat consumption was 230 million tonnes. The forecast for 2050 is 465 million tonnes. There is a strong relationship between meat consumption and rising per person incomes (Figure 6.15), although anomalies

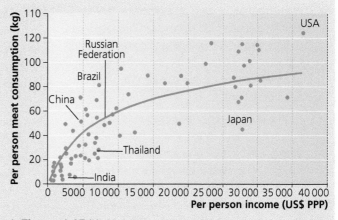

▲ **Figure 6.15** Meat consumption and income

do occur due to cultural traditions.

It is no coincidence that many committed environmentalists are vegetarian. A study at the University of Chicago calculated that changing from the average American diet to a vegetarian one could cut annual emissions by almost 1.5 tonnes of carbon dioxide.

Arriving at an evidenced conclusion

Although there are many uncertainties in projections of climate change, all evidence points to increased global temperatures. Since the beginning of the industrial era, the temperature has already risen by 0.8°C, and some reports (Reuters, 2014) suggest that predicted emissions from power plants, factories and cars have locked the globe on a path towards an average temperature rise of almost 1.5°C above pre-industrial times by 2050, with other studies predicting that the world will have emitted enough carbon dioxide to warm the planet 2°C by about 2036 (Le Page, 2015).

Moreover, we cannot be certain what will happen due to the complexities of carbon cycling, storage and feedback mechanisms. The decisions made at the Paris accord (page 157), however, offer humanity the mechanisms to alter behaviour and change course away from the more extreme predictions of climate change. The worst impacts of global warming could be avoided by cutting greenhouse gas emissions and enhancing mitigation strategies. Altering our diets to lower meat consumption, using renewable sources of energy and living more sustainable lifestyles could result in lower levels of warming, towards the lower estimates given by the IPCC (Figure 1.15). It seems that the future very much lies in our hands.

Chapter summary

✔ Decarbonisation refers to a large reduction of carbon dioxide per value of gross world product. Since most of the carbon dioxide comes from burning fossil fuels, a sharp reduction in the use of fossil fuels or a large-scale system to capture and sequester the carbon dioxide is needed.

✔ A decarbonised economy is based on power sources that either do not use carbon-based fuels or limit their use. Economies that are decoupled from fossil fuels have, in theory, a minimal output of power-related carbon dioxide emissions into the atmosphere.

✔ Eventually, non-renewable resources could become completely exhausted. Renewable energy can be used over and over again. The sources of renewable energy are mainly forces of nature that are sustainable and that usually cause little or no environmental pollution. Renewable energy includes hydro-electric (HEP), biomass, wind, solar, geothermal, tidal and wave power.

✔ There are, however, potential geographic challenges linked with most renewable energy sources, ranging from biodiversity loss to world trade rules and issues.

✔ Technological fixes such as carbon capture and storage may prove to be humanity's best bet. However, urgent action is needed now to develop new technologies on the scale necessary to prevent a high-emissions scenario.

✔ Anthropogenic climate change poses a serious threat to the health of the planet. There is a range of adaptation and mitigation strategies that could be used, but for them to be successful they require global agreements as well as national actions. As a result, most experts fear a warmer world is now inevitable. The question which remains to be answered is: *How much warmer will it get?*

Refresher questions

1   Outline the difference between climate change mitigation and adaptation strategies.

2   Using examples, explain what is meant by global governance in relation to carbon cycle management.

3   Explain how renewable energy could be used to help decarbonise economic activity at local and global scales.

4   Using examples, outline geographical issues that may prevent some renewable energy sources from becoming more widely used.

5   Explain how carbon capture and storage can be used to reduce levels of atmospheric carbon dioxide.

6   Outline the role of the United Nations REDD Programme in reducing atmospheric carbon dioxide concentrations.

7   Suggest how the resilience and ability of a country to adapt to climate change depends on its: geographic position; population density; economic development status; the local availability of natural resources.

8   Outline (i) how ice cores are used to measure atmospheric carbon dioxide and methane concentrations over the last 800,000 years, and (ii) what the evidence shows.

9   Explain the 'hockey stick' trend line indicated by atmospheric carbon dioxide concentrations measured from Vostok ice cores and recent Mauna Loa data.

Discussion activities

1   In small groups, share your prior knowledge about the energy mix of your home country. What percentage is derived from oil, gas and coal? How important are renewables in the overall mix? How has the mix changed over time? Alternatively, work in pairs to research this information.

2   In pairs, discuss possible synoptic linkages between the changing energy mix in a country such as the UK and other geography topics, including the water cycle and changing places. For example, the development of HEP impacts on the water cycle. The gradual movement away from the use of coal in the twentieth century led to the deindustrialisation of many mining towns. What were the consequences of this for people and the environment?

3   In small groups, draw a mind map showing all of the complexities and uncertainties surrounding the issue of climate change and the extent to which it can be prevented. Try to draw knowledge from as many different geography topics – including global systems and global governance – as possible.

4   In pairs, develop a list of different mitigation strategies that can be adopted by players and stakeholders representing different geographical scales, including local, city, national, international (EU) and global. Discuss the relative importance of these different scales of action for successful mitigation.

5   Hold a whole class debate about the likelihood that significant global warming will have been prevented during your lifetime. The class can divide into two teams – optimists and pessimists. Both teams should work together to create the best argument they can, drawing on as many different geographical ideas, theories and pieces of information as possible. Hold a class debate, addressing the question: 'To what extent is a warmer world inevitable?' Divide into two teams, each addressing a different side of the argument.

FIELDWORK FOCUS

The global scale of the geographic themes and issues covered in this chapter are not always suitable for an A-level Geography independent investigation requiring you to collect primary data. However, some students may live and work in a particular local context where renewable energy and climate change mitigation initiatives are being carried out, and where opportunities for primary data collection exist. For example:

A *An investigation of the impacts of a small scale renewable energy scheme.* Primary data collection opportunities may include interviews with local people or scheme managers focusing on possible land-use conflict or other issues. It would be possible to collect quantitative data in the case of afforestation or peatland restoration. For example, measuring the carbon storage of a new forest can give an indication of the role of trees in reducing carbon dioxide concentrations in the atmosphere. The following site has ideas to get you started: www.forestry.gov.uk/pdf/Revised-biomass-equations-27Jan2014.pdf/$FILE/Revised-biomass-equations-27Jan2014.pdf

B *An investigation of carbon footprint management in a local place.* A range of primary and secondary data could be collected, including: personal actions to reduce carbon emissions (well-designed questionnaire and population sampling strategy would be required); actions by businesses, both local stores and chain stores such as Tesco (interviews with managers could be carried out; social and environmental responsibility reports are published annually by large chain stores); research into local authority actions (including recycling schemes) and the way that these are informed by national, EU and global laws or agreements.

Further reading

Dickie, A., *et al.* (2014) *Strategies for Mitigating Climate Change in Agriculture: Abridged Report.* California Environmental Associates, www.climatefocus.com/sites/default/files/strategies_for_mitigating_climate_change_in_agriculture.pdf

Intergovernmental Panel on Climate Change, www.ipcc.ch

Le Page, M. (2015) 'The Climate Fact No One Will Admit: 2°C Warming Is Inevitable', *New Scientist,* www.newscientist.com/article/dn28430-the-climate-fact-no-one-will-admit-2-c-warming-is-inevitable

Reuters (2014) 'Some Climate Change Impacts Unavoidable – World Bank'. www.reuters.com/article/climatechange-impacts/some-climate-change-impacts-unavoidable-world-bank-idUSL6N0TD0DA20141123

CHAPTER 7

Carbon and water cycle interrelationships

The carbon cycle and the water cycle are essential both for human existence and the functioning of ecosystems we depend on. They provide many ecosystem services that benefit humanity but are vulnerable to human impacts too. This chapter:

- explores local-scale carbon and water interrelationships
- examines carbon and water relationships in the tropical rainforest
- investigates feedback between and within water and carbon cycles in the Arctic tundra.

KEY CONCEPTS

Positive feedback A change in a system which triggers further changes, leading in turn to even greater deviation from the original system state.

Ecosystem services The benefits provided by ecosystems, including supporting services, e.g. primary productivity, recycling of nutrients, soil formation; regulating services, e.g. climate-water regulation, pollination, production of goods; provisioning services, e.g. services obtained from ecosystems including peat, water, fish; cultural services, e.g. recreation, improvements to human health, spiritual wellbeing.

① Investigating local-scale carbon and water interrelationships

▶ *How are carbon and water cycling affected by local-scale interrelationships?*

Throughout this book, significant interactions between water cycling and carbon cycling have been highlighted, often at a very local scale. This has included consideration of how:

- the two cycles interact directly when carbon is transported in solution by river water (see page 109)
- ecosystems function as stores of both water and carbon, and influence the way a range of carbon transfers and water flows operate (see page 119)
- deforestation can lead to increased overland flow, which in turn causes soil erosion and the permanent loss of carbon storage capacity (see page 56)
- increasing atmospheric carbon concentrations are changing global climate and water cycle flows and stores (see page 140).

Additional integrated case studies of land use can be used to illustrate links between water and carbon cycling at the local scale. For example, the process of desertification serves to emphasise:

● how water and carbon cycles are closely coupled
● how water cycle changes are linked with soil and vegetation loss
● the interdependence of soil and vegetation health.

For example, Figure 7.1 shows how desertification in a developing world region such as the Sahel is driven partly by climate change (or fluctuations) and also by land-use change. A series of interlinked changes affect the carbon and water cycle as follows:

● Reduced vegetation cover may lead to reduced carbon sequestration in biomass (due to lower photosynthetic fixation of carbon) and also reduced soil carbon.
● Less vegetation cover may reduce the remaining soil's infiltration capacity; this means infiltration-excess runoff will take-place when rain does occur (see page 40), resulting in soil erosion and gullying.
● Accelerated soil erosion will reduce soil carbon storage even more.
● With less soil cover, ecosystem net primary productivity (involving biomass carbon sequestration and storage, see pages 111–12) will fall further.

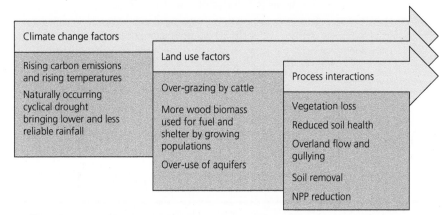

▲ **Figure 7.1** How water and carbon cycle changes are linked when desertification is taking place. Source: Simon Oakes, (2017), *Geography Study and Revision Guide: Global Change*, Hodder

Studying local cycle linkages in the UK

Significant interactions between water cycling and carbon cycling can be explored at the local scale. Floodplains, parks and even small gardens or patches of wasteland can provide a study context. For example, in autumn and winter each year you can observe fallen leaves being washed by heavy rainfall into gutters and sewers. In rural and urban drainage basins alike, high levels of overland flow and channel flow within the water cycle are responsible for transporting a large volume of carbon (stored as leaf litter biomass and soil organic matter) away from the land.

Figure 7.2 shows organic litter in Epping Forest. Several geographical questions relevant to this chapter's focus on cycle interrelationships are generated by this image.

We might ask: what role will precipitation play in helping this litter to decompose? What role could overland flow play in transporting this litter to other places, or into a stream? In the longer term, what role will this litter play in the water cycle if it remains here, decomposes and becomes stored as humus in the soil?

▲ **Figure 7.2** Forest floor litter in Epping Forest

Deforestation and the water and carbon cycles

▲ **Figure 7.3** Felling of timber greatly accelerates water flows through a drainage basin; moreover, overland flows may carry large amounts of carbon-rich tree debris into streams and rivers

Chapter 2 examined the impacts of forest removal on hydrological systems. This is a land-use change which in fact involves a series of interactions between local water and carbon cycles which can be studied in relatively localised areas of the UK, for example wherever a tract of a Forestry Commission plantation is felled to produce timber (Figure 7.3). Deforestation brings significant disruption to the water balance and to carbon cycles.

- Runoff processes are invariably affected by vegetation removal and ground surface disturbance. For example, with a complete forest canopy, interception is high, infiltration high, overland flow limited and therefore soil erosion rates are low.
- Once the canopy is removed, raindrop impact on the surface may increase, dislodging fine particles and blocking soil pores. This results in reduced infiltration and increased overland flow now that neither interception nor infiltration are significant influences on water storage and flow patterns.

Coppicing Cutting trees down to a level near the ground surface so as to encourage new growth. This traditional woodland management strategy is still used today in many forests in lowland Britain, including Slapton National Nature Reserve, Devon and Epping Forest (located within the London commuter belt).

Pollarding Refers to cutting trees down to a height of about 2 metres or so. This is above the level which browsing animals, such as deer and goats, can reach on their hind feet. Local authorities in the UK use pollarding to manage the growth of trees lining streets in urban areas; typically, trees are cut every ten to fifteen years.

a

b

▲ **Figure 7.4 (a)** Coppicing **(b)** Pollarding

The methods used to clear forest can have a major impact on soil loss. Compaction by machines can increase the amount of sediment removal by further reducing infiltration. Increased overland flow, sheetwash and gullying may then erode large amounts of soil.

Sustainable management of local forest resources

One way of minimising carbon and water cycle disruption is to adopt sustainable practices of **coppicing** and **pollarding**. Trees are managed, rather than removed, meaning that hydrological changes and any subsequent soil erosion will be minimised. This also means fewer significant changes for local carbon cycling and storage.

Coppicing refers to the cutting down of trees to ground level every few years or so (Figure 7.4). This fosters the vigorous regrowth of new and relatively regular-sized shoots (which are therefore more useful economically). The renewed growth of trees also increases carbon sequestration without any significant long-term changes in local hydrological cycle operations.

Coppiced woodlands are generally harvested in sections on a rotation basis. This means that there will be sections of woodland present which are of varying ages, thereby enhancing the potential for biodiversity locally. Coppiced woodlands are very different in appearance from natural forests. They have an open, sunny feel with relatively few large trees. Without too much shade, coppice regrowth (and carbon sequestration) can occur quickly. In coppiced woods and forests, broad access tracks are usually found, which are used to extract the harvested wood shoots each winter.

 # Amazonian carbon and water cycles: connections and changes

▶ *What interrelationships exist between changing carbon and water cycles in Amazonia?*

The second part of this chapter provides an in-depth case study of the Amazonian tropical rainforest biome in order to illustrate and analyse key themes in water and carbon cycles, their interrelationships and changes in both systems that are being driven by environmental change and human activity. Amazonia is of particular significance given the key role that its own carbon and water stores and cycles play in supporting life on Earth, especially in regard to climatic regulation.

However, some computer models predict a large-scale substitution of Amazon forest by savanna-like vegetation by the end of the twenty-first

AQA

OCR

WJEC/Eduqas

century as the result of a combination of local, regional and global pressures and changes. This forest dieback is concerning:

- The Amazonian biosphere contains 90–140 billion tons of carbon, equivalent to approximately 90 to 140 years' worth of current global anthropogenic carbon emissions.
- Amazonia plays a vital role in the global water cycle too. Approximately 8 trillion tons of water evaporates from Amazon forests each year. The rest of the rainfall entering the Amazon basin flows into the Atlantic Ocean, accounting for around 15 to 20 per cent of the worldwide continental freshwater runoff to the oceans.
- Also, the Amazon is home to approximately one out of every five mammal, fish, bird and tree species in the world.

> **KEY TERM**
>
> **Forest dieback** A condition when a large number of trees are killed due to environmental conditions such as drought, high temperatures, pollution (e.g. acidification) or diseases.

Carbon storage uncertainties in Amazonia

Amazonia is an incredibly important carbon store (at over 5.5 million km^2, it is roughly half the size of Europe). As they grow, the Amazon rainforest's 3 billion trees account for one-quarter of the carbon dioxide absorbed by the land each year. In theory, as anthropogenic carbon dioxide emissions increase (see Chapter 5), forests will absorb and store more carbon, provided they have enough water and nutrients to grow.

However, some academics believe that the Amazon has already passed a threshold in terms of how much surplus carbon it can store.

Between 1983 and 2011, a team of scientists measured trees in over 300 plots, recording their number, size and density to calculate how much carbon each one stored. They found that the trees took up carbon and grew more quickly during the 1990s, before levelling off after 2000.

Biomass mortality has actually *increased* since 1985. The combination of relatively constant productivity and increasing tree mortality suggests that the carbon storage in the Amazon has actually *declined* since the 1990s. This begs the question: why are some trees dying?

One explanation is that the faster some trees grow (in response to a greater abundance of carbon dioxide), the sooner they reach maturity and age. Also, as taller trees are more vulnerable to high winds and drought, faster growth (on account of enhanced greenhouse gas emissions and greater atmospheric carbon storage) may be putting trees at increased risk from weather extremes. During droughts in Amazonia in 2005 and 2010, researchers found that there were increased peaks of tree mortality, albeit short-lasting. However, the overall trend of increased tree mortality had begun *before* these drought periods (nevertheless, some researchers still believe that drought may be a possible underlying reason for declining carbon storage).

- Predictions of a continuing increase of carbon storage in tropical forests may therefore be misplaced. Instead, increased biomass mortality may mean that carbon is being released, rather than sequestered, by forests.
- This is most unfortunate because it had been hoped that increased carbon sequestration by forests might be a natural negative feedback effect that could help return the global carbon cycle to a state of equilibrium again.
- It is not known whether the trend of declining Amazonian carbon storage is replicated more widely across other tropical forests in central Africa and Indonesia. If it is indeed happening elsewhere, the outlook for the world's climate is not good.

Local water and carbon cycle changes in Amazonia driven by human activity

Agriculture and land use conversion

There are mounting human pressures to convert parts of the Amazon rainforest to arable and pastoral farmland, resulting in further reductions in terrestrial carbon storage and water cycle interception storage.

- Since the eradication of foot-and-mouth disease in large areas of southern and eastern Amazonia, the region's cattle and pig industries have been expanded.
- Secondly, the increasing international demand for agro-industrial commodities has coincided with a scarcity of appropriate land in the USA, Western Europe and China.
- Thirdly, the rising price for oil has stimulated the expansion of the biofuels industry. The Brazilian sugar cane industry is one of the world's most efficient and inexpensive forms of ethanol, and palm oil is one of the most efficient sources of biodiesel. Figure 7.6 shows a process-response system diagram (see Chapter 1) which synoptically links changes in global economic systems with land use and carbon cycle responses.
- Finally, genetic modification has produced crops that are tolerant of the heat and humidity associated with rainforest climates. All of these have put pressure on the Amazon to produce more agricultural products for a growing world market.

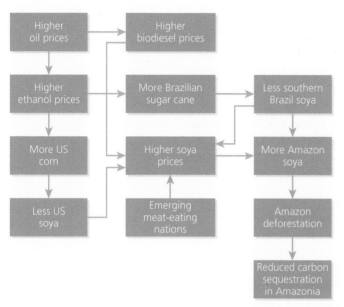

▲ **Figure 7.5** Interactions between changing global economic systems (commodity prices) and carbon and water cycle changes (Amazon deforestation)

Growing global demands for biofuels and animal feed provide new incentives to clear forest that coincided with a decade-long expansion of the Amazon cattle herd. Most of the expansion of soya production has been onto areas that were previously cattle pasture. However, the displaced cattle ranchers are, as a result, pushing even deeper into pristine primary forest regions. In the Brazilian North region, over 70 million head of cattle occupy 84 per cent of the total area under agricultural and livestock uses, and have expanded 9 per cent yr^{-1}, on average, causing 70–80 per cent of deforestation in this region, over a period of ten years.

Damage done by fire, fragmentation and logging

Parts of the Amazonia region experience a dry season (defined as any month with less than 100 millimetres of rain) which generally lasts from July to September. In fact, nearly half of the forests of the Amazon maintain full leaf canopies during dry seasons that range from three to five months, indicating a high tolerance of drought. These forests have many trees with deep roots and are able to avoid severe drought stress by absorbing moisture stored deep in the soil. However, even the Amazon forest's drought tolerance has its limits.

Land-use activities increase forest susceptibility to fire during periods of drought by providing ignition sources, by fragmenting the forest and by thinning the forest through logging. Although ignition from lightning is rare, man-made sources of ignition are increasingly abundant. Forests are burned in preparation for crops or pasture, and to improve pasture forage. However, these fires frequently burn beyond their intended boundaries into neighbouring forests. During the severe drought of 1998, approximately 39,000 km² of standing forest caught fire, which is twice the area that was clear-cut that year. During the severe drought of 2005, at least 3000 km² of standing forest burned in the southwest Amazon. These low, slow-moving fires can kill up to 50 per cent of trees (above 10 cm in diameter). Forest fire can increase susceptibility to further burning in a positive feedback by killing trees, opening the canopy and enabling increased solar radiation to reach the forest floor.

- Forest degradation is also fostered by fragmentation and edge effects associated with forest clear-cutting for pasture formation. This is because tree mortality and forest flammability are higher along forest edges.
- Tree mortality induced by drought, fire, fragmentation and selective logging is the first step of a process of forest degradation that is reinforced by positive feedbacks that can, potentially, convert forest ecosystems into fire-prone 'brush' vegetation.
- In Mato Grosso and south-eastern and eastern-Para states, repeated burning has converted the Amazon into a secondary forest dominated by invasive species, grasses, bamboo and ferns.

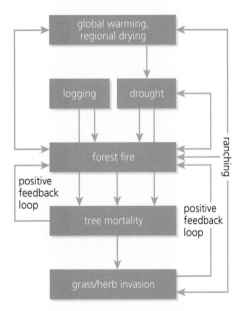

▲ **Figure 7.6** How different interactions and positive feedback processes can result in forest being replaced with fire-tolerant (pyrophytic) grasslands

ANALYSIS AND INTERPRETATION

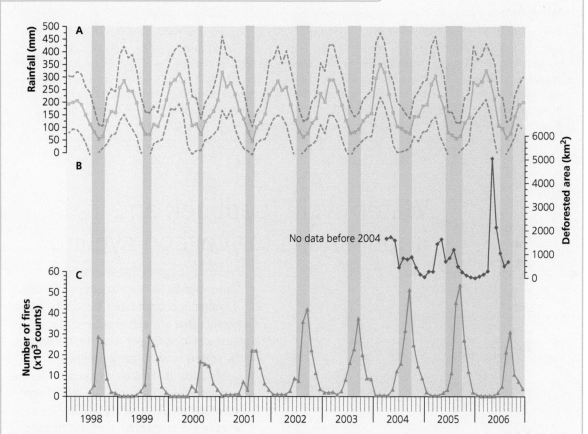

▲ **Figure 7.7 (a)** Mean rainfall (mm), **(b)** cumulative deforested area (km²), and **(c)** cumulative number of fires in part of Amazonia. Fires release a lot of carbon from the biosphere to the atmosphere

(a) Study Figure 7.7. Compare the seasonal timing of the most severe fires with that of deforestation.

GUIDANCE

Most of the severe fires occur towards the end of the dry season, e.g. 2002, 2004 and 2005. An exception is 1998 when the most severe fire occurred during the middle of the dry season. In contrast, most deforestation tends to occur towards the end of the wet season, i.e. about three months after the peak of the wet season, prior to the start of the dry season.

(b) Explain why fires and deforestation may occur at different times.

GUIDANCE

Deforestation occurs before the dry season, as the cut wood can be left to dry out before burning. Farmers set fire to the wood towards the end of the dry season as it will burn more easily, and they can clear the land for farming. In addition, burning fertilises the soil in time for the new growth of crops that are planted.

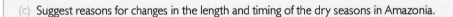

(c) Suggest reasons for changes in the length and timing of the dry seasons in Amazonia.

Water cycle feedback and its implications for carbon cycling

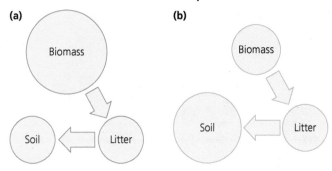

▲ **Figure 7.8** The proportionate size of carbon-containing nutrient stores in **(a)** tropical rainforest and **(b)** tropical grasslands

Evidence suggests that the eastern Amazon may be drier in future as a result of system feedback (see Chapter 1). It now appears that any clearing beyond approximately 30 per cent of the forests of the region may trigger a decline in rainfall. Rainfall in the region may decline by as much as 20 per cent by 2100. Some scientists suggest that there could be a large-scale, late-century substitution of closed-canopy evergreen forest by savanna-like and semi-arid vegetation, mostly in the eastern part of the basin.

This is because markedly lower evapotranspiration rates (and also the higher albedo of savanna vegetation compared with dark rainforest) reinforces the drying through a positive feedback loop. If water flows from biomass and land to the atmosphere become reduced, there is less water vapour present to condense and form precipitation (see page 121).

This is a positive feedback loop which, in theory, may lead to further grassland colonisation. This has implications for carbon cycling: Figure 7.8 compares carbon storage and cycling in forest and grassland biomes. As you can see, they differ greatly with proportionately far less carbon stored as living biomass in the savanna.

Water budget deficits for varying timescales

As we have already learned, water cycle deficits (see pages 64–7) may become more common in places where there are fewer trees because of reduced evapotranspiration rates. This would most likely represent a long-term and permanent decline in water availability.

There are other reasons for water budget deficits on varying timescales:

- Cyclical occurrence of drought in parts of Amazonia is linked with changes in sea surface temperatures along the Pacific coast of South America. If they increase, rainfall decreases in the Amazon. This typically happens during El Niño events (see page 17). Studies show tree mortality in central Amazonia increasing from around 1 to 2 per cent in El Niño years. Reductions in NPP and carbon storage also occur at such times, mainly due to changes in soil moisture and light availability. There is some debate among scientists whether these drought cycles will become more frequent or severe in the future on account of global warming.
- Changing water availability can also occur over a shorter timescale due to the outbreak of fire. This can lead to reduced rainfall. Anecdotal evidence from Mato Grosso suggests that the rainy season begins later in the year by several weeks when the density of smoke is high.

Implications of water budget deficit for carbon storage and cycling

In one Amazonian study, researchers found that tree mortality increased during drought periods, and above ground biomass declined by as much as 25 per cent. Total wood production was decreased by around 12 per cent, and the leaf area index also declined. This reduced trees' ability to photosynthesise. However, wood production increased rapidly following the end of the drought. This is partly due to reduced competition for light, moisture and minerals among the surviving trees.

The intense droughts that are predicted for the Amazon could lead to increased release of carbon to the atmosphere. Even short-term drought can reduce wood production, and so carbon storage in the forest. If this continues, it may become a carbon source rather than a carbon sink.

Climate change and forest change interactions

The climate and air quality in Amazonia depend strongly on the character of the vegetation cover. Any large-scale removal of vegetation cover might therefore be expected to in turn modify the local climate while also contributing to global climate change through the release of stored carbon – possibly leading to reductions in precipitation in Amazonia, which might ultimately trigger a cascade of biosphere changes, including effects on flora, fauna and soils! As you can see, there are many possible interactions and interlinkages to critically examine.

IPCC scientists acknowledge that the possible future relationship between atmospheric warming and changes in precipitation patterns is highly uncertain (see page 15). A number of climate models do suggest, however, that global warming could lead to lower rainfall for Amazonia. The potential warming and drying of the Amazonian climate could lead to a dieback of large areas of forest. Relative to bare soil, vegetation (especially forest) can enhance evaporation, precipitation and infiltration.

- Forest loss leads to less water being available locally (less interception, more overland runoff, resulting in there being less water to evaporate). Thus, although there is more heating, there may also be less convectional rainfall because so much water has flowed out of the area as runoff.
- Forest loss also increases surface albedo, which reduces convection, causing a further positive feedback on rainfall reduction.

KEY TERM

Hadley Cell An atmospheric circulation cell whereby air rises near the equatorial low pressure zone, rises to the tropopause and flows polewards before sinking at the sub-tropical high pressure belt around 20–30° north and south, and some of the air returning equatorwards.

The positioning of Amazonia on the Equator means that large-scale forest loss could ultimately exert far-reaching and wide-scale effects by modifying the global atmospheric circulation. With forest present, high rates of evaporation cause a larger proportion of the energy to be transferred to the atmosphere and form the basis of the Hadley convection cell circulation. However, with forest removal, there is less latent heat input into the upper atmosphere, and consequently the Hadley Cell is weakened.

Amazonian climatic drying and forest dieback is certainly a complex coupled process, as Figure 7.9 shows. The drying is related to changes in sea surface temperatures (SSTs). Despite the potential for enhanced biomass

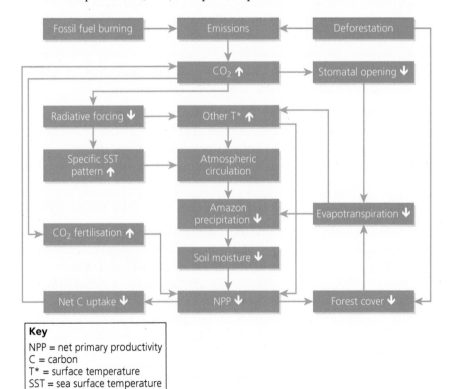

Key
NPP = net primary productivity
C = carbon
T* = surface temperature
SST = sea surface temperature

▲ **Figure 7.9** Potential feedback mechanisms involved in Amazonian climate change and forest degradation

growth that increased atmospheric CO_2 brings, the warming and drying of the climate is predicted to cause forest dieback. Two positive feedback loops contribute to the projected precipitation reduction in some models: (1) reduced forest cover causes a reduction in evaporation, and (2) carbon release caused by dieback contributes to the global CO_2 rise, which further accelerates global warming and regional precipitation changes. Approximately one-third of a recent precipitation reduction in western Amazonia has been attributed to these two feedback cycles.

Towards a tipping point

Economic, ecological and climatic systems in Amazonia may now be interacting in ways which will move the forest beyond a threshold, or tipping point. Multiple human and physical processes are involved, including forest clearance and thinning, fragmentation, fires and period droughts linked with El Niño events. Global warming further reinforces many of these processes by increasing air temperatures, dry season severity and the frequency of extreme weather events.

One projection for 2030 suggests that 31 per cent of the Amazon closed-canopy forest formation will be deforested and 24 per cent will be damaged by drought or logging (Figure 7.10). If 'business as usual' continues, logging will further reduce the Amazonian forest carbon store size by 15 per cent; drought damage will cause another 10 per cent reduction in forest biomass. The effects of more frequent fires could be to release an additional 20 per cent of forest carbon to the atmosphere. Thus, some predicted 15–26 Pg of carbon contained in the Amazon will be released to the atmosphere in less than 30 years as a result of large-scale dieback.

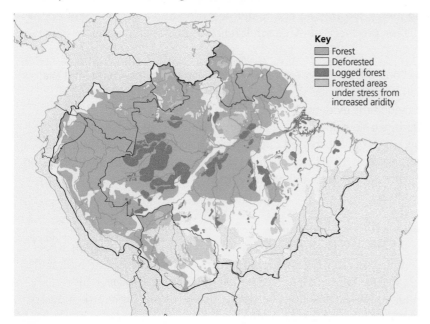

Key
- Forest
- Deforested
- Logged forest
- Forested areas under stress from increased aridity

▲ **Figure 7.10** The Amazon rainforest, 2030, showing drought-damaged, logged and cleared forest assuming 'business as usual'

Protecting Amazonia: strategies, actions and players

Given the essential role Amazonia plays in regulating carbon and water cycle dynamics – at both local and global scales – action is urgently needed to prevent to prevent further forest biomass loss. This section briefly reviews three governance interventions involving government and non-government players acting at both national and global scales: these are the soya moratorium, the cattle moratorium and the UN-REDD programme.

1 The soya moratorium

In the twentieth century, deforestation was seen as necessary for development and modernisation in South America. That attitude appears to have changed, and now it is seen as a wasteful and exploitative destruction of resources which deliver vital ecosystem services. A number of environmental pressure groups have helped change public opinion and political will both in Amazonian producer countries and in countries where timber or agricultural products from deforested land are consumed.

The first dramatic change came in the soybean industry. In 2006, Greenpeace released *Eating up the Amazon*, a report that showed the connections between the soybean industry and deforestation, global warming, water pollution and sometimes slave labour. It focused on two multinational companies, Cargill and McDonald's. Within weeks, the soy industry declared an end to deforestation and a commitment not to buy beans produced on Amazon lands that had been deforested since 2006.

- By 2009–10, less than 1 per cent of soybeans were grown on land that had been deforested since 2006. The use of remote sensing data provided the evidence of the moratorium's success.
- Despite a slow down in the amount of land being cleared for farming, soy yields have increased due to multiple cropping. For example, the harvest for 2013–14 was about 95 million tonnes, up 7 million tonnes on the previous year. Brazil has now become the world's largest soybean producer.

2 The cattle moratorium

After soy, the next largest direct cause of Amazonian deforestation is beef cattle industries. In 2009, two NGO reports – Friends of the Earth (Brazil)'s *Time to Pay the Bill* and Greenpeace's *Slaughtering the Amazon* – drew attention to inescapable links between the expansion of the cattle industry and destruction of Amazonian forest.

AQA

OCR

WJEC/Eduqas

- To manage the situation, slaughterhouses around the world agreed to buy cattle only from ranchers registered with the Brazilian rural environmental land registry. For example, Brazil's biggest domestic beef buyers, Wal-Mart, Carrefour and Pão de Açúcar, suspended contracts with suppliers involved in Amazon deforestation.
- In order to register, ranchers needed to provide the GPS co-ordinates of their property, thereby enabling a comparison of ranch locations with areas undergoing deforestation.
- The cattle moratorium has forced change, but it happened more slowly than with soy.

3 The UN-REDD global governance programme

The UN-REDD Programme is the United Nations Initiative on Reducing Emissions from Deforestation and Forest Degradation (REDD) in low-income countries. This global governance programme was launched in 2008 and involves the Food and Agriculture Organisation (FAO), the United Nations Development Programme (UNDP) and the United Nations Environment Programme (UNEP). In 2007, it was estimated that the forests, through deforestation, forest degradation and land-use changes, contributed approximately 17 per cent of global greenhouse gases (see also page 171).

The UN-REDD Programme supports national-scale actions in over 50 countries, in Africa, Asia Pacific and Latin America. REDD stresses the role of conservation, the sustainable management of forests and the increase of forest carbon stocks. By June 2014, total funding had reached almost US$200 million; Norway is the leading donor country.

The Brazil–Norway agreement is the largest programme overseen by REDD anywhere in the world, and it has already provided US$670 million in compensation for the reductions made in the first few years after its signing. The Amazon forest is assumed to contain 100 tonnes of carbon per hectare (although it is probably higher over much of the region), and the estimated reduction in emissions is paid for at a fixed rate of five dollars per ton of CO_2. However, the importance of the agreement with Norway is political and symbolic, not just financial.

Future prospects for greater protection

The net effect of the soya and cattle moratoria – combined with REDD agreements – is illustrated in Figure 7.11. This shows a reduction in annual deforestation rates since around 2003–04.

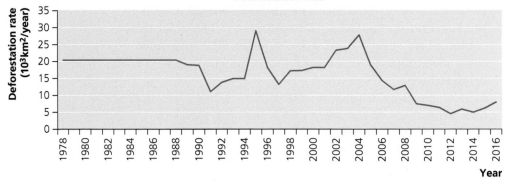

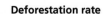

▲ **Figure 7.11** Annual deforestation rates in Amazonia (pre-1989 figures are estimates)

Some recent developments throw doubt on whether Amazonia's future can still be secured, however. Data for Brazil show a 28 per cent increase in the 2012–13 deforestation rate compared with 2011–12. Although the amount of forest clearance was still 70 per cent below the 1996–2005 average, this upward swing is deeply concerning. Since then it has increased, somewhat erratically (but overall the rate of deforestation is much lower than in the 1980s and 1990s). The greatest forest clearance continues to be mainly in the states of Para and Mato Grosso, where most of Brazil's agricultural expansion is focused.

Satellite data show that over 6600 km² of forest were destroyed between August 2016 and July 2017, compared to nearly 7900 km² in 2015/2016, a decline of 16 per cent over one year. The rate remains well above the 4571 km² deforested in 2012, which was the low since records began in 2004/2005. Reasons for the decline included stepped-up enforcement and real-time monitoring that allow for rapid response to deforestation. However, economic recession in Brazil and a drop in livestock prices were also likely to be responsible for the decline.

Until 2012, Brazil seemed on track to meet its international climate change commitment of zero illegal deforestation by 2030 and a national target of reducing deforestation to no more than 3900 km² a year by 2020. Through tough enforcement, Brazil slashed its deforestation rate by 83 per cent to 4571 km² a year between 2004 and 2012. However, it may prove difficult for Brazil to manage to reduce deforestation to 3900 km² by 2020 in order to meet climate change targets. Brazil's supreme court has upheld major changes to laws that protect the Amazon, reducing penalties for past illegal deforestation, frustrating environmentalists trying to protect the world's largest rainforest.

③ Investigating the Arctic tundra

▶ *In what ways is the Arctic tundra changing?*

Cold tundra climates are found in areas of high latitude and altitude, especially inland areas far from the warming influence of the sea. Covering 20 per cent of the Earth's land surface (Figure 7.12), tundra environments are located adjacent to glacial regions; they are not directly beneath ice but do experience very cold weather for most of the year, leaving the ground beneath permanently frozen. For example, much of the US state of Alaska is a tundra region where there is complete darkness for up to one month of the year and temperatures may fall to minus 50°C.

This combination of low temperatures (and thus low air humidity) and permafrost means that, for the most part, tundra environments are characterised by slow rates of water and carbon cycling when viewed over longer timescales. Decomposition occurs so slowly in the Arctic tundra that recognisable plant and animal remains may lie on the surface of the ground for many months, or even years. Photosynthesis does not take place for large parts of the year due to climatic limiting factors. However, there is a sudden rise in biological productivity levels each year during the short two- or three-month growing season.

Similarly, the melting of snow in spring or early summer can cause the sudden release of large flows of water previously held in cryosphere storage. Page 82 explored the regime of the River Colorado: the snowmelt which brings a rapid rise in discharge during April and May takes place in tundra regions within the river's enormous catchment. Approximately one-third of the Rocky Mountains reach altitudes above the limit where trees may grow in northern Colorado. Starting at elevations of around 3500 metres, these are Alpine tundra environments where snow collects during winter each year. Tundra is largely norther peatland, and is a major store of carbon.

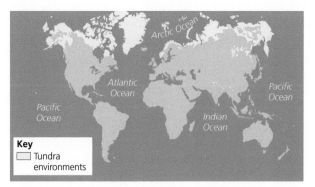

▲ **Figure 7.12** The global distribution of tundra environments

Projected impacts of climate change on tundra water and carbon cycling

The presence of permafrost in tundra environments means that they form part of the cryosphere, as Chapter 1 explained, along with ice sheets, ice caps, glaciers, permafrost and seasonal snow and ice cover. The cryosphere is a water store which is particularly vulnerable to global climate change; and tundra permafrost environments are, in turn, widely viewed as the most vulnerable part of the cryosphere. This is because cyclical seasonal thawing and refreezing of the frozen topsoil naturally occurs; however, any rise in mean annual temperature is likely to increase both the amount of thawing and the length of thawing each year, prior to refreezing in winter, if that still happens at all. Under these changing conditions, water losses via evaporation and runoff are projected to increase, resulting in a permanent loss of cryosphere water storage.

Most models of global climate change suggest that warming will be highest at high latitudes, especially during winter months.

In the Fennoscandian region (Norway, Sweden, Finland and the Kola Peninsula of Russia) a doubling of atmospheric CO_2 levels could lead to an increase of mean winter temperatures by 5–6°C, mean annual temperatures by 4–5°C, an increase in the length of the growing season by 70 to 150 days and annual precipitation to increase by 150–300 mm. These will lead to radical changes in the distribution of tundra environments. In Canada, for example, the southern limit of permafrost will shift northwards by between 100 and 200 kilometres.

In addition, there are many positive feedback mechanisms which can accelerate the increase of temperature in tundra environments. Melting sea ice and reduced snow cover reduces the albedo (reflectivity) of the surface, and increases solar radiation input into the ground and the sea. In boreal forests and wetlands, frozen soils contain large reserves of carbon and methane (CH_4) that are released as ground temperatures rise and permafrost thaws.

The tundra environment is therefore highly vulnerable to significant change on account of global climate change. The corollary of this is that changes in tundra environments may, in turn, accelerate global climate changes because of large-scale carbon transfers from the ground to the atmosphere.

Accelerated environmental changes in tundra areas

The Intergovernmental Panel on Climate Change (IPCC) predicts that land surfaces will warm more rapidly than oceans, and that high latitudes will warm more than low latitudes, especially in winter. Based on an eventual doubling of atmospheric CO_2 (to a level of 550 ppm or greater), some scientists predict that winter temperatures may rise by 8 to 12°C, and mean temperatures by 1.5 to 4.5°C by 2050. These changes are up to five times greater than those seen during the whole of the twentieth century. Such developments would have a profound impact on the whole of the global climate system, due to feedback mechanisms.

For example, increased tundra biomass production and decay, as well as the decomposition of organic matter previously frozen in the permafrost, would transfer vast quantities of CO_2 and methane in the atmosphere, leading in turn to further atmospheric warming and greater precipitation in tundra regions. Local moisture availability may rise further due to increased melting of permafrost, rising ground temperatures and hence reduced freezing of lakes and rivers (thereby leading to increased evaporation rates).

Other possible impacts on water and carbon cycling may include the following:

- *Glacier surges in adjacent polar regions.* Sometimes glaciers naturally advance, especially after years of heavy snowfall. This can cause tundra grasses to be covered, thereby changing patterns of both cryosphere and biomass storage.
- *Bursting lakes.* Catastrophic changes occur when a glacial meltwater lake bursts. Summer meltwater running off a glacier sometimes builds up behind a natural barrier, such as a ridge or bank of soil. Failure of the barrier can release a surge of meltwater, destroying local biomass while also radically altering local water storage and flow patterns.
- *Solifluction.* Tundra regions are vulnerable to seasonal soil melting. This can trigger a process caused solifluction. Rather like a landslide, whole sections of a slope start to move under gravity. The surface vegetation gets rolled beneath the moving mass of soil, like the tracks on a tank, resulting in transfers of organic carbon-rich material from one place to another.

In these and other ways, accelerated local water cycling is to be expected alongside faster carbon cycling.

Degrees of uncertainty

Future projections for climatic and environmental change involve great complexity and uncertainty, however. For example, it is not always easy to identify longer-term changes when short-term seasonal fluctuations are happening (this relates to the concept of a dynamic or changing equilibrium explored in Chapter 1). Moreover, Arctic tundra data are variable in coverage, quality, type and reliability. Nevertheless, many studies have suggested that ground temperatures have increased significantly since the 1970s, and in many locations there is clear evidence of permafrost melting and retreating. For example, in the European Arctic and Russian subarctic, a marked rise in the temperature of the top 3 metres of permafrost has been recorded. Similar results have been found in northern Canada and the Qinghai-Xizang (Tibet) Plateau.

▲ **Figure 7.13** The Arctic tundra contains many wetlands, and is a major source or carbon and methane

However, tundra environments are varied in character and it is important not to generalise too much about potential impacts of atmospheric warming. For example, you may think that higher spring temperatures would most likely lead to more rapid snow melting. Yet if winters become slightly warmer the air becomes able to hold more moisture which could lead to an increase in winter snowfall – meaning there is more snow to melt in spring! The net impact on cryosphere storage is therefore uncertain. Moreover, these different impacts (more snow melting or more snow falling) would have contrasting effects on ground albedo. If warming leads to reduced snow cover and albedo decreases, the result will be further atmospheric warming, i.e. positive feedback. On the other hand, if warming leads to increased snow cover (due to higher precipitation in the form of snow), surface albedo will increase, leading to a reduction in warming of the air, i.e. negative feedback.

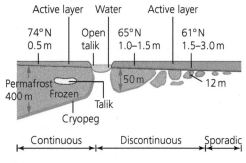

▲ **Figure 7.14** Types of permafrost

Changing permafrost projections

As we have seen, permafrost plays a very important role in water and carbon cycling and storage at both global and local scales. There is uncertainty over how fast melting will occur in the future. An increase in ground temperature can certainly lead to a reduction of permafrost thickness, or even its elimination in areas where the ice is naturally quite shallow or sporadic (Figure 7.14). A more general concern is that thawing permafrost will release methane, leading to a positive feedback of increased warming, melting of permafrost and increased release of methane.

Rising temperatures will have a major impact on the thermal regime of permafrost. With a rise of temperature, the active layer increases (thickens), and there is increased melting of the permafrost both at its top and its base. Whereas the increase in the depth of the active layer changes almost immediately, the melting of the permafrost at its base may take hundreds or thousands of years, as soil and rock are poor conductors of heat. For example, studies in northern Alaska have shown that increased surface temperatures of 4°C led to an increase of the active layer from 50 cm to 93 cm, but that at a depth of 30 m temperatures had increased by only 1°C and at 100 m there was no temperature change.

CONTEMPORARY CASE STUDY: INTERLINKED CARBON AND WATER CYCLE CHANGES IN THE SIBERIAN TUNDRA

The Siberia Integrated Regional Study (SIRS) aims to investigate environmental change in Siberia (Figure 7.15) under the current environment of global change, and the potential impact on Earth system dynamics. The actual and projected consequences of global warming are well documented for Siberia. Temperatures have already increased, particularly in the winter in Eastern Siberia (0.5°/decade), and the number of frost days and growing season length have also increased (roughly one day yr[-1]).

Future climatic change threats include the shift of permafrost boundaries northward, dramatic changes in land cover and the entire hydrological regime of the territory. These processes feed back to, and influence, climate dynamics through the exchange of energy, water, greenhouse gases and aerosols. However, scientists have limited precise knowledge about the processes that control physical change in this region. They anticipate a range of possible and potentially unexpected feedbacks to and from terrestrial and aquatic systems.

According to SIRS, there are three main regional challenges which are very important for the global carbon cycle:

■ Permafrost degradation, especially its border shift, might form a significant carbon and methane source to the atmosphere (while also seriously threatening human infrastructure). Climate-related drying would alter biogenic emissions in peatlands that have been deposited over millennia and would increase the potential for peat fires which cannot be extinguished.

■ Temperature/precipitation/hydrology regime changes, which might increase risks of peat fires, thus changing significantly the carbon, terrestrial and hydrologic cycle of the region.

■ Shifts in ecosystem borders northwards, which will also change regional input into the global carbon and radiation balance.

Future biosphere-climate system interactions in northern Eurasia

Three climatic variables – growing degree days (GDD5, the sum of daily temperatures above 5°C), negative degree days (the sum of daily temperatures below 0°C) and annual moisture index (the ratio between GDD5 and annual precipitation) – were used as inputs to the Siberian bioclimatic model (in order to generate a pattern of ecosystem distribution for 2009 and 2090 climate (Figure 7.16). Dramatic predicted changes in land cover by 2090 include the following:

■ Tundra and the forest–tundra ecotone zones (currently one-third of the Siberian area) will practically disappear.

▲ **Figure 7.15** Siberian tundra

- The taiga zones (currently two-thirds of Siberia) will move northward and be reduced to 40 per cent of the present area.

- In the northern latitudes and highlands, tundra would be replaced by forest (with decreased albedo thus triggering further changes) in 28 per cent of the territory.

The southern borderline of taiga in Siberia will be more affected by forest fires. In this region, conditions favourable to fire have become unusually frequent during the past two decades. Fire and the melting of permafrost are considered to be the principal mechanisms that will trigger vegetation changes across the Siberian landscape.

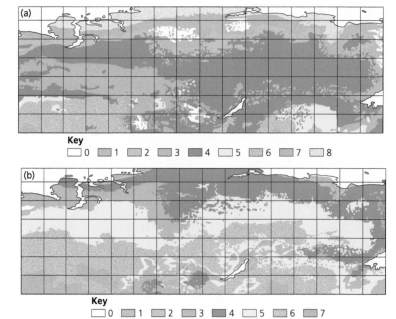

Key

☐ 0 ▦ 1 ▦ 2 ▦ 3 ▦ 4 ☐ 5 ▦ 6 ▦ 7 ☐ 8

Key

☐ 0 ▦ 1 ▦ 2 ▦ 3 ▦ 4 ☐ 5 ▦ 6 ▦ 7

Key

0 Water
1 Tundra
2 Forest-tundra
3 Dark-leaf taiga
4 Light-leaf taiga
5 Forest-steppe
6 Steppe
7 Semi-desert
8 Polar desert

▲ **Figure 7.16** Vegetation over Siberia (**a**) 2009, and (**b**) predicted for 2090

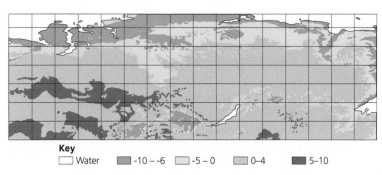

Key ☐ Water ■ -10 – -6 ☐ -5 – 0 ☐ 0–4 ■ 5–10

▲ **Figure 7.17** Albedo change in Siberia by 2090 (%) caused by changes in vegetation cover in a warming climate

This book has demonstrated how the water and carbon cycles are complex, interlinked systems upon which all life on Earth ultimately depends. The predicted future changes for Siberia and Amazonia which have featured in this chapter give important insights into the implications of future climate change for our planet's 'life support systems' – and, in turn, for the individuals, societies and places that depend on these systems and the ecosystem services they help provide.

Chapter summary

✔ The operations of the water cycle and the carbon cycle are closely interrelated, both at local scales (e.g. a UK woodland or local places in the Sahel affected by desertification) and continental scales (Amazonia and the Siberian or Alaskan tundra).

✔ Positive feedback effects linking water and carbon cycle changes in the Sahel have brought sometimes irreversible environmental changes as part of the desertification process.

✔ Deforestation radically alters carbon and water storage and flow patterns; coppicing and pollarding are forest management practices that can help limit changes in both systems.

✔ Amazonia is a globally important carbon and water store that is under great pressure due to drought, forest fires and developmental changes in South America. There is uncertainty surrounding interactions between climate change and forest change in Amazonia, particularly in relation to the rainforest's capacity to serve as a 'sink' for anthropogenic carbon emissions.

✔ Players at a range of scales have taken action to help limit the destruction of Amazonia; the UN-REDD global governance programme is especially important.

✔ Climate change is predicted to have a major impact on the world's high latitudes. Arctic tundra environments are especially at risk, since they experience seasonal changes between frozen and unfrozen ground conditions. Warming of this world region has major implications for global climate due to the way albedo changes and feedback mechanisms may potentially accelerate change even further.

Refresher questions

1. Outline how water and carbon cycle interactions can contribute to the process of desertification.
2. Explain how the practices of coppicing and pollarding allow forests to be used as a resource while also minimising water and carbon cycle disruption.
3. Explain why rainforest trees may suffer reduced carbon storage capacity as a result of higher atmospheric temperatures.
4. Outline the main physical and human pressures on the Amazonian rainforest.
5. Explain (i) how the loss of tropical rainforest can lead to the occurrence of drought, (ii) the possible role of feedback processes in forest dieback.
6. Define the following terms: permafrost; tundra.
7. Explain why water and carbon cycling and tundra regions naturally occur very slowly except at certain times of the year.
8. Explain two different positive feedback mechanisms involving water and carbon cycling that could occur in tundra environments as a result of global climate change.

Discussion activities

1. In pairs, discuss possible interrelationships of carbon and water in your local area. Possible contexts to think about include a school playing field; areas of parks and woodlands; the gardens of local homes.
2. Consider also how it may be possible to protect these local water and carbon cycles and their interrelationships.
3. In small groups, compare the roles played by tropical rainforest and Arctic tundra in global water and carbon cycling. Which is most important and why? Does the loss of tropical rainforest or degradation of the Arctic tundra pose the greatest threat to life on Earth, and why?
4. Other than tropical rainforest and Arctic tundra, what are the other main global biomes? In small groups choose one of these other biomes and discuss ways in which carbon and water cycling are interlinked in this biome. For example, you might choose deciduous forest or savanna grassland.

FIELDWORK FOCUS

The topic of the interrelationships between carbon cycles and water cycles could be used as a basis for an A-level individual investigation. However, because you would need to research two different cycles – and additionally investigate links between then – this would be quite a challenging topic to undertake.

A *Investigating links between the water and carbon cycles at the local scale.* One possible topic might be estimating how carbon is being moved in solution by a stream. This can be studied over different timescales through repeated visits at different times of the year.

Alternatively, you might just take a 'snapshot' set of readings on one particular day. It might be advisable to seek help from a chemistry teacher at your school: they will be able to offer advice on the best way to analyse water content. You would need to independently devise a robust sampling strategy.

B *Investigating links between water flows and leaf litter cycling.* Another topic could be studying autumn leaf fall in your local area and the subsequent recycling or removal of these organic remains. Water flows – particularly overland flow – could play an important role. It might be possible to devise a strategy for measuring the amount of organic debris carried in a small stream. What techniques and sampling strategies could you use?

Further reading

Aragao, L., *et al.* (2008) 'Interactions between rainfall, deforestation and fires during recent years in the Brazilian Amazonia', *Philosophical Transactions of the Royal Society*, B, 363, pages 1779–85

Betts, R., *et al.* (2008) 'Effects of large-scale Amazon forest degradation on climate and air quality through fluxes of carbon dioxide, water, energy, mineral dust and isoprene', *Philosophical Transactions of the Royal Society*, B, 363, pages 1873–80

Brando, P., *et al.* (2008) 'Drought effects on litterfall, wood production and belowground carbon cycling in an Amazon forest: results of a throughfall reduction experiment', *Philosophical Transactions of the Royal Society*, B, 363, pages 1839–48

Brienen, R., *et al.* (2015) 'Long-term decline of the Amazon carbon sink', *Nature* 519, pages 344–8

French, H. (2017) *The Periglacial Environment*, Wiley-Blackwell

Gordov, E., and Vaganov, E. (2010) 'Siberia integrated regional study: multidisciplinary investigations of the dynamic relationship between the Siberian environment and global climate change', *Environmental Research Letters*, 5, 015007 (5 pages)

Lloyd, J., and Farquhar, G. (2008) 'Effects of rising temperatures and CO_2 on the physiology of tropical forest trees', *Philosophical Transactions of the Royal Society*, B, 363, pages 1737–46

Nepstad, D., *et al.* (2008) 'Interactions among Amazon land use, forests and climate: prospects for a near-term forest tipping point', *Philosophical Transactions of the Royal Society*, B, 363, pages 1811–17

Vygodskaya, N. N., *et al.* (2007) 'Ecosystem and climate interaction in the boreal zone of northern Eurasia', *Environmental Research Letters*, 2, 045033 (7 pages)

World Wildlife Fund, WWF Russia (2015) 'Summary comments to the Environmental and Social Impact Assessment of the project for construction of integrated complex for extraction and liquefaction of natural gas and gas condensate on the Yamal peninsula (Yamal LNG Project)'

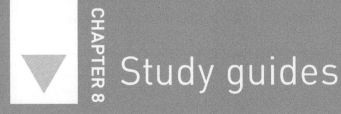
① AQA A-level geography: Water and carbon cycles

Content guidance

The compulsory topic of Water and carbon cycles (Topic 3.1.1) is supported fully by this book. As part of the assessment, students are expected to apply a wide range of geographical skills including data manipulation and statistics, along with evaluative essay-writing.

Sub-theme and content	Using this book
3.1.1.1 Water and carbon cycles as natural systems This section introduces the key concepts and terminology required for the study of systems in physical geography. Important ideas include system inputs and outputs, feedback mechanisms and equilibrium.	Chapter 1, pages 1–24
3.1.1.2 The water cycle Water cycle characteristics and processes are studied at different spatial scales and over different timescales. ■ Firstly, the global pattern of water storage must be understood, including the lithosphere, hydrosphere, cryosphere and atmosphere. ■ The processes responsible for global and localised changes over time in these stores include: evaporation, condensation, cloud formation, precipitation and cryospheric changes. ■ The drainage basin is studied as an open system whose inputs and outputs are influenced by a wide range of hydrological processes such as interception and overland flow. ■ The water balance, flood hydrographs and (natural and human) influences on water cycling over time.	Chapter 2, pages 25–64 Chapter 3, pages 65–98
3.1.1.3 The carbon cycle Carbon cycle characteristics and processes are studied at different spatial scales and over different timescales. ■ Firstly, the global pattern of carbon storage must be understood, including the lithosphere, hydrosphere, cryosphere, biosphere and atmosphere. ■ The factors and processes affecting changing patterns of storage include: biological processes including photosynthesis; carbon sequestration in oceans and sediments; and weathering.	Chapter 4, pages 99–101; 102–110

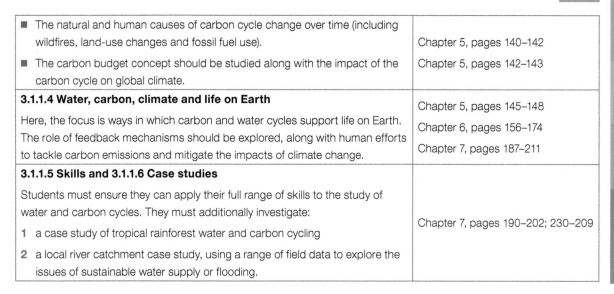

The natural and human causes of carbon cycle change over time (including wildfires, land-use changes and fossil fuel use).	Chapter 5, pages 140–142
The carbon budget concept should be studied along with the impact of the carbon cycle on global climate.	Chapter 5, pages 142–143
3.1.1.4 Water, carbon, climate and life on Earth Here, the focus is ways in which carbon and water cycles support life on Earth. The role of feedback mechanisms should be explored, along with human efforts to tackle carbon emissions and mitigate the impacts of climate change.	Chapter 5, pages 145–148 Chapter 6, pages 156–174 Chapter 7, pages 187–211
3.1.1.5 Skills and 3.1.1.6 Case studies Students must ensure they can apply their full range of skills to the study of water and carbon cycles. They must additionally investigate: 1 a case study of tropical rainforest water and carbon cycling 2 a local river catchment case study, using a range of field data to explore the issues of sustainable water supply or flooding.	Chapter 7, pages 190–202; 230–209

AQA assessment guidance

Water and carbon cycles are assessed as part of Paper 1 (7037/1). This examination is 2 hours and 30 minutes in duration and has a total mark of 120. There are 36 marks allocated to Water and carbon cycles. This consists of:

- a series of three short-answer questions (worth 16 marks in total)
- one 20-mark evaluative essay.

Water and carbon cycle short-answer questions (up to 6 marks)

Your first Water and carbon cycles question should be a knowledge-based short-answer task worth 4 marks and targeted at assessment objective 1 (AO1 – knowledge and understanding) using the command term 'explain'. High marks will be awarded to students who can write concise, detailed answers which link together a range of ideas, concepts or theories. As a general rule, try to ensure that every point is either developed or exemplified:

- a developed point takes the explanation a step further (for example, by adding extra detail of how a process operates)
- an exemplified point refers to a relatively detailed or real-world example in order to support the explanation with evidence.

Your second short-answer question will make use of a resource (map/diagram/table) and is targeted at assessment objective 3 (AO3 – quantitative, qualitative and fieldwork skills). It will ask you to analyse or extract meaningful information or evidence from the information provided. The question will most likely use the command terms 'analyse', 'compare' or 'assess'. The analysis and interpretation features included throughout this book help you develop the skills you need to answer this type of question successfully.

Your third and final short-answer question will again make use of a figure but is now targeted mainly at assessment objective 2 (AO2 – *application* of knowledge and understanding). It will use a command phrase such as 'Analyse the figure and using your own knowledge...' You are therefore expected to use the data as a springboard to apply your own geographical ideas and information. For example, a 6-mark question could accompany a graph showing the potential impacts of global climate change: 'Using the figure and your own knowledge, assess the relative importance of physical and human causes of carbon budget changes.' You can answer using your own knowledge rather than extracting detailed information from the resource.

Water and carbon cycles evaluative essay writing

The 20-mark Water and carbon cycles essay will most likely use a command word or phrase such as 'to what extent', 'how far', 'assess the extent' or 'discuss'. The mark scheme will be weighted equally towards AO1 and AO2. For instance:

- **'Human activity has caused irreversible damage to the interrelationships between the carbon cycle and the water cycle.' To what extent do you agree with this view?**

The box below provides guidance on how to answer this type of question.

Writing an evaluative essay about the water and/or carbon cycles

Chapters 1–6 of this book contain a section called Evaluating the issue. These have been designed specifically to support the development of evaluative essay-writing skills. As you read each Evaluating the issue section, pay particular attention to the way:

- *Underlying assumptions and possible contexts are identified at the outset.* Before planning your answer think very carefully about what kinds of contrasting contexts you could choose to write about (arid or non-arid environments; rainforest or tundra environments; places affected by ENSO cycles in varying ways; small or large drainage basins; surface water or groundwater stores, etc.).
- *Your essay can be carefully structured around different paragraphed themes, views, concepts or scales of analysis.* Often, essay questions will ask you to discuss or evaluate the 'role', 'significance', 'importance' or 'benefits' of something (in relation to different system flows or stores, for example). Consider this exam-style essay question: 'Assess the severity of the impact of drought on different system stores.' Before answering, ask yourself: what timescales could be used when responding? Could the threshold concept be applied, perhaps by suggesting some changes might be irreversible? Important questions such as these should be thought about at the planning stage of your essay and may help form an introduction.

Command words and phrases such as 'evaluate', 'to what extent' and 'discuss' require you to reach a final judgement. Don't just sit on the fence. Draw on all the arguments and facts you have already presented in the main body of the essay, weigh up the entirety of your evidence and say whether – on balance – you mostly agree or disagree with the question you were asked. To guide you, here are three simple rules.

I   *Never sit on the fence completely.* Essay titles are created purposely to generate a discussion which invites a final judgement following debate. For example, the question: 'To what extent are human impacts on the water cycle always negative?' Do not expect to receive a really high mark if you end your response with a phrase such as: 'So all in all, some human impacts on the water cycle process are negative while others are not.'

2 *Equally, it is best to avoid extreme agreement or disagreement.* In particular, you should not begin your essay by dismissing one viewpoint entirely, for example by writing: 'In my view, all human impacts on the water cycle turn out negative eventually and this essay will explain all of the reasons why this is the case.' It is essential that you consider a range of arguments or different points of view.

3 *An 'agree, but…' or 'disagree, but…' judgement is usually the best position to take.* This is a mature viewpoint which demonstrates you're able to take a stand on an issue while remaining mindful of other views and perspectives.

Synoptic geography

In addition to the three main AOs, some of your marks are awarded for 'synopticity'. The box below explains what this means.

Thinking synoptically

Instead of focusing on one isolated topic, you are expected to draw together information and ideas from across your specification. You will be making connections between different 'domains' of knowledge, especially links between people and the environment (i.e. connections across human geography and physical geography). The study of Yemen's water crisis (page 000) is a good example of synoptic geography because of the important linkages between water cycle management and political issues; so too is the study of desertification (page 000) because it links together economic development with water cycling dynamics.

Throughout your course, take careful note of synoptic themes whenever they emerge in teaching, learning and reading. Examples of synoptic themes include: the impact of place changes on water cycle flows and stores; strategies to reduce carbon footprint sizes of different places in order to help mitigate anthropogenic emissions into the carbon cycle; the role of globalisation in changing the global carbon cycle. Whenever you finish reading a chapter in this book, make a careful note of any synoptic themes that have emerged (they may have been highlighted, or these could be linkages that you work out for yourself).

AQA's synoptic assessment

Some 9- or 20-mark exam questions may require you to link together knowledge and ideas from different topics. These may appear in both your physical geography and human geography examination papers. For example, a water cycle question (Paper 1) might ask you to think about ways the characteristics of different places could affect water and carbon cycles:

- **How far do you agree that globalisation is responsible for harmful changes to water and carbon cycles?**
- **How far do you agree that predicted carbon cycle changes will have negligible impact on the operation of physical landscape systems?**

For the second question, the mark scheme would include the following statement: 'This question requires links to be made across the specification specifically between Water and carbon cycles and content drawn from Coastal/Glacial/Hot desert systems and landscapes.' One way to tackle this kind of potentially tricky question is to draw a mind map when planning your response. Draw two equally sized circles and fill these with relevant ideas, processes and contexts, trying to achieve the best balance you can between the two linked topics.

Edexcel A-level geography: Physical systems and sustainability

Content guidance

The compulsory topics of *The water cycle and water insecurity* (Topic 5) and *The carbon cycle and energy security* (Topic 6) are supported fully by this book.

Topic 5: The water cycle and water insecurity

Enquiry question and content	Using this book
1 What are the processes operating within the hydrological cycle from global to local scale? The topic begins by exploring the global hydrological cycle's operation as a closed system composed of different water stores and the fluxes which connect them. Consideration is given to the different residence times of these stores. Additionally, the water cycle is studied at a more localised level. The drainage basin is an open system whose operation depends on many processes (such as precipitation, infiltration, overland flow, etc.). Physical and human factors determine the relative importance of different system elements (for example, groundwater abstraction). Finally, this section examines local-scale water budgets and storm hydrographs in addition to larger-scale river regimes. A range of supporting examples is required, including a study of the role of planners in managing drainage basin land use.	Chapter 1, pages 1–24 Chapter 2, pages 25–64
2 What factors influence the hydrological system over short- and long-term timescales? There are a range of reasons for deficits within the hydrological cycle, including drought, ENSO cycles and misuse of surface and groundwater resources (the impacts of drought on ecosystems should also be explored). Surpluses within the hydrological cycle should be studied too – causes include storms, monsoon rainfall and human actions that increase flood risk. A study should be carried out of the environmental damage and economic impacts caused by flooding. Finally, climate change impacts on the hydrological cycle and global water stores must be studied. These changes may be short-term (ENSO) or longer-term (future projections for global warming).	Chapter 2, pages 25–64
3 How does water insecurity occur and why is it becoming such a global issue for the 21st century? This begins with an analysis of the physical and human causes of water insecurity. You should be aware of global patterns of water stress and water scarcity. Important issues include salt water encroachment, over-abstraction from water stores, pollution and increasing human demands leading to rising water insecurity and projections of future water scarcity. The consequences of water insecurity should also be understood including their implications for economic activity, human health and potential conflict amongst users both within and between different countries (especially in relation to trans-boundary water sources). Finally, the sustainability of different water management approaches must be considered. This includes hard engineering schemes (e.g. dams), water conservation measures, integrated drainage basin management for large rivers and international frameworks (e.g. Water Framework Directive).	Chapter 3, pages 65–98

Topic 6: The carbon cycle and energy security

Enquiry question and content	Using this book
1 How does the carbon cycle operate to maintain planetary health? This section introduces the global carbon cycle, its stores and annual fluxes. Attention is paid to the importance of geological storage and processes. Biological processes are another important focus including ocean organisms, terrestrial ecosystems and carbon stored in soils. In addition, this section looks at how the carbon cycle sustains other Earth systems and regulates the atmosphere while also maintaining soil health. Fossil fuel combustion, however, has modified the carbon cycle with implications for other physical systems.	Chapter 4, pages 99–117 Chapter 5, pages 140–143 Chapter 7, pages 187–211
2 What are the consequences for people and the environment of our increasing demand for energy? A range of energy security issues are investigated, including national energy consumption and energy mix. The factors affecting access to energy resources are looked at alongside the main energy players such as TNCs and OPEC. This section also focuses on the link between fossil fuel use and economic development: important ideas include energy pathways linking areas of fossil fuel production and consumption; and the drive to develop unconventional fossil fuel resources such as shale gas. A case study is required, for example different attitudes to fracking in the USA. Finally, alternative energy sources such as nuclear and wind power must be assessed in terms of their costs and benefits. Biofuels and new technologies (e.g. carbon capture and storage) should also be studied.	Chapter 5, pages 126–139 Chapter 6, pages 159–169; 173–174
3 How are the carbon and water cycles linked to the global climate system? A range of human activities and issues threaten the carbon and water cycle. These include deforestation and other land-use changes, ocean acidification because of fossil fuel combustion and climate change (with a focus on the frequency of drought and the health of forests as carbon stores). In turn, this section explores the implications for humans of forest loss and changing water and carbon cycles. Particular issues include: changing precipitation patterns, river regimes and global water stores; and deteriorating ocean health leading to marine resource losses. This part of the course concludes with a look at the potential for large-scale carbon cycle changes occurring due to feedback mechanisms such as permafrost melting. A range of adaptation and mitigation strategies should be investigated at both national and global scales, along with varying attitudes of citizens and states about the need for action.	Chapter 5, pages 145–148 Chapter 6, pages 156–179 Chapter 7, pages 187–211

Edexcel assessment guidance

Water and carbon cycles are assessed as part of 9GEO/01. This examination is 2 hours and 15 minutes in duration and has a total mark of 105. There are 49 marks allocated to Physical systems and sustainability (Topic 5: The water cycle and water insecurity and Topic 6: The carbon cycle and energy security). This consists of:

- a series of three short-answer questions (worth 17 marks in total)
- one 12-mark evaluative 'mini-essay'
- one 20-mark evaluative essay.

Water and carbon cycle short-answer questions

Your first short-answer questions will make use of a resource (map/diagram/table) and is targeted partly at assessment objective 2 (AO2 – involving the analysis or extraction of meaningful evidence from the information provided). It will most likely use the command term 'explain'. The analysis and interpretation features included throughout this book are intended to support the study skills you need to answer this kind of question successfully.

Two further questions – worth 6 marks and 8 marks respectively – will be knowledge-based tasks targeted at assessment objective 1 (AO1 – knowledge and understanding). Expect use of the command term 'explain'. High marks will be awarded to students who can write concise, detailed answers which incorporate a range of ideas, concepts or theories. As a general rule, try to ensure that every point is either developed or exemplified:

- A developed point takes the explanation a step further (for example, by adding extra detail of how a process operates).
- An exemplified point refers to a relatively detailed or real-world example in order to support the explanation with evidence.

Evaluative essay writing

The 12-mark task (this makes use of stimulus material) and the 20-mark essay will most likely use the command words 'assess' and 'evaluate' respectively. Both will have a mark scheme which is weighted heavily towards AO2. For instance:

- **Assess the likely impacts of climate warming on the components of the carbon cycle shown in the figure. (12 marks)**
- **Evaluate the extent to which today's increasing demand for energy is the main reason for contemporary carbon cycle changes. (20 marks)**

The box on pages 214–215 provide guidance on how to answer these questions.

Synoptic geography

In addition to the three main AOs, some of your marks are awarded for 'synopticity'. The box on page 215 explains what this means.

Edexcel's synoptic assessment

Synoptic exam questions are worth plenty of marks and you need to be well prepared for them.

- In the Edexcel course, an entire examination paper is devoted to synopticity: Paper 3 (2 hours 15 minutes) is a synoptic 'decision-making' investigation. It consists of an extended series of data analysis, short-answer tasks and evaluative essays (based on a previously unseen resource booklet).
- As part of your Paper 3 answers, you will need to apply a range of knowledge from different topics you have learned about and also make good analytical use of the previously unseen resource booklet (the 'analysis and interpretation' features in this book have been carefully designed to help you). The context used in the resource booklet may well make use of themes drawn from *The water cycle and water insecurity* (Topic 5) and *The carbon cycle and energy security* (Topic 6).

OCR A-level geography: Earth's life support systems

Content guidance

The compulsory topic of Earth's life support systems (Topic 1.2) is supported fully by this book. The detailed content is structured around four sub-themes.

Enquiry question and content	Using this book
1 How important are water and carbon to life on Earth? This introductory section examines the importance of water and carbon in supporting life on Earth and human activity. Students are expected to gain an understanding of system cycling and the difference between open and closed systems. The main inputs, outputs and stores of both systems must be examined in detail, along with the range of physical processes and pathways that operate (such as evapotranspiration, photosynthesis, overland flow, infiltration, carbon sequestration, etc.).	Chapter 1, pages 1–24 Chapter 2, pages 25–64 Chapter 4, pages 99–117
2 How do the water and carbon cycles operate in contrasting locations? Two key case studies are used in this section of the specification: ● Firstly, study is required of the physical and human factors that affect the water and carbon cycles in the tropical rainforest. Changes over time should be studied including those caused by human land-use modifications. There should be consideration of strategies to restore equilibrium to rainforest, water and carbon cycles. ● The second case study is water and carbon cycling in the Arctic tundra. Physical and human factors affecting the cycles include climate, vegetation, relief features, seasonal changes and the human impact of the energy industry. Management strategies should also be investigated.	Chapter 7, pages 187–211
3 How much change occurs over time in the water and carbon cycles? Two key ideas feature in this section. First, human factors – ranging from land-use changes and water extraction to fossil fuel combustion – can disturb the equilibrium of water and carbon cycles. The importance of positive and negative feedback loops should also be understood. Second, natural changes over different timescales must be appreciated. These include seasonal short-term changes to the cycles and long-term changes over millions of years. As part of this, you should explore how changes are researched and monitored, and why.	Chapter 1, pages 1–24 Chapter 2, pages 25–64 Chapter 3, pages 65–98 Chapter 4, pages 114–123 Chapter 4, pages 140–141 Chapter 5, pages 146–148

Enquiry question and content	Using this book
4 To what extent are the water and carbon cycles linked?	
You must be familiar with the concept of interdependency: in the context of this topic, this means the way the water and carbon cycles are linked via ocean, atmospheric and other processes. Human activities, in particular those resulting in long-term climate change, cause changes to both linked cycles. Finally, study should be undertaken of global management strategies to protect both the carbon cycle and the water cycle. Actions include afforestation, wetland restoration, carbon trading agreements, improved water management and drainage basin planning.	Chapter 3, pages 65–98 Chapter 6, pages 156–161; 171–172 Chapter 7, pages 187–211

OCR assessment guidance

Water and carbon cycles are assessed as part of Paper 1 (H481/01). This examination is 1 hours and 30 minutes in duration and has a total mark of 66. There are 33 marks allocated to Water and carbon cycles. This consists of:

- a series of three short-answer questions (worth 17 marks in total)
- one 16-mark evaluative essay.

Earth's life support systems short- and medium-length questions

Your first short-answer questions will make use of a resource (map/diagram/table) and is targeted at assessment objective 2 (AO2 – involving the analysis or extraction of meaningful evidence from the information provided). It will most likely use the command term 'suggest'. The analysis and interpretation features included throughout this book are intended to support the study skills you need to answer this kind of question successfully.

The second, worth 3 marks, will be targeted at assessment objective 3 (AO3 – geographical skills and techniques). There may be the need to carry out some mathematical operation such as completing a statistical test analysing some data.

Finally, a medium-length 10-mark question, most likely using a command word such as 'examine', will be targeted jointly at AO2 and assessment objective 1 (AO1 – knowledge and understanding). High marks will be awarded to students who can write concise, detailed answers which incorporate a range of ideas, concepts or theories. As a general rule, try to ensure that every point is either developed or exemplified:

- A developed point takes the explanation a step further (for example, by adding extra detail of how a process operates).
- An exemplified point refers to a relatively detailed or real-world example in order to support the explanation with evidence.

Evaluative essay writing

The 16-mark essay will most likely use a command word or phrase such as 'discuss', 'assess' or 'how far do you agree'. The mark scheme will be weighted equally towards AO1 and AO2. For instance:

- **'Human factors affect the water cycle more significantly in the tropical rainforest than in the Arctic tundra.' Discuss.**

The box on page 217 provides guidance on how to answer this type of question.

Synoptic geography

In addition to the three main AOs, some of your marks are awarded for 'synopticity'. The box on page 218 explains what this means.

OCR's synoptic assessment

In the OCR course, part of Paper 3 (Geographical debates H481/03) is devoted to synopticity. For this exam, you will have studied two optional topics chosen from: Climate change, Disease dilemmas, Exploring oceans, Future of food and Hazardous Earth. In Section B of Paper 3, you must answer two synoptic essays worth 12 marks each.

Each synoptic essay links together the chosen option with a topic from the core of the A-level course, such as Changing spaces; Making places. Possible Paper 3 essay titles might therefore include:

- **'Changes in the water and carbon cycles are the greatest threat to food security.' Discuss.**
- **'The hazards related to water and carbon cycles could become more important in future than geo-physical hazards.' Discuss.**

One way to tackle these kinds of questions would be to draw a mind map to help plan your response. Draw two equally sized circles and fill these with relevant ideas, processes and contexts, trying to achieve the best balance you can between the two linked topics. The mark scheme requires that your answer includes: 'clear and explicit attempts to make appropriate synoptic links between content from different parts of the course of study'.

WJEC and Eduqas A-level geography: Global systems (water and carbon cycles)

Content guidance

Both WJEC and Eduqas students must study the compulsory topic of Global systems: water and carbon cycles (WJEC Topic 3.1; Eduqas Topic 2.1) which is supported fully by this book. When aiming for the highest grades, study of processes and contexts should focus on:

- the extent to which systems are in a state of equilibrium or experience positive feedback processes (pages 11–15)
- the different temporal scales over which flow and storage patterns change
- ways in which carbon and water cycle operations are linked and interdependent (Chapter 7).

Enquiry question and content	Using this book
1 The concepts of system and mass balance	
This introductory section explores the concepts of system and mass balance, and the importance of temporal and spatial scales in relation to system dynamics.	Chapter 1, pages 1–24
2 Catchment hydrology – the drainage basin as a system	
This section deals with catchment hydrology, systematically exploring different inputs, flows, stores and outputs of drainage basins.	Chapter 2, pages 25–64

Enquiry question and content	Using this book
3 Temporal variations in river discharge Here, the focus is on the characteristics and factors affecting river regimes, the climatic and local catchment factors influencing the shape of hydrographs.	Chapter 2, pages 49–54
4 Precipitation and excess runoff within the water cycle Key processes of air uplift, condensation, cloud formation and precipitation formation. This section also deals with climatic and human causes of excess runoff generation.	Chapter 2, pages 39–40, 46–48
5 Deficit within the water cycle This section deals with causes of water cycle deficits, ranging from meteorological explanations across varying timescales, and human interventions including aquifer modifications.	Chapter 3, pages 64–67
6 The global carbon cycle You need to be able to describe and explain the operation of the global carbon cycle, including the chemical, biological and physical processes responsible for carbon sequestration and flow pathways.	Chapter 4, pages 99–117
7 Carbon stores in different biomes This section deals with carbon cycling and storage in two case study biomes, the tropical rainforest and temperate grassland; human intervention must also be studied.	Chapter 4, pages 118–123 Chapter 7, pages 187–211
8 Changing carbon stores in peatlands over time Here, the focus shifts to peat formation, and the way human activity can reduce or restore carbon storage in peat environments.	Chapter 4, pages 103, 121 Chapter 6, pages 159–160
9 Links between the water and carbon cycles This section explores anthropogenic carbon emissions and the implications of increasing atmospheric carbon for water cycle movements and Earth's oceans. There is also a requirement to briefly study local water and carbon cycle linkages.	Chapter 7, pages 187–211
10 Feedback within and between the carbon and water cycles This last section reviews the operation of positive and negative feedback loops applied to the context of cryosphere, marine and terrestrial carbon storage. The implications for life on Earth of permafrost melting are particularly important.	Chapter 5, pages 145–148 Chapter 7, pages 187–211

WJEC and Eduqas assessment guidance

Water and carbon cycles is assessed as part of:

- *WJEC Unit 3.* This examination is 2 hours in duration, and has a total mark of 96. There are 35 marks allocated for Water and carbon cycles, indicating that you should spend around 45 minutes answering. The 35 marks consists of:
  - two structured short-answer questions (together worth 17 marks)
  - one 18-mark evaluative essay (from a choice of two).

- *Eduqas Component 2.* This examination is 2 hours in duration, and has a total mark of 110. There are 40 marks allocated for Water and carbon cycles, indicating that you should spend around 45 minutes answering. The 40 marks consists of:
  - two structured short-answer questions (together worth 20 marks)
  - one 20-mark evaluative essay (from a choice of two).

Both courses use broadly similar assessment models and these are dealt with jointly below.

Short-answer questions
WJEC
Short-answer questions 1 and 2 on your examination paper include several different types of short-answer question.

Part (a) of one question – *but not the other* – will usually be targeted at assessment objective 3 (AO3) and is worth 3 marks. This means that you will be required to use geographical skills (AO3) to analyse or extract meaningful information or evidence from the figure. These questions will most likely use the command words 'describe', 'analyse' or 'compare'.

In another short-answer question, worth 5 marks, you may be asked to apply your knowledge and understanding of water and carbon cycles in an unexpected way. This is called an applied knowledge task; it is targeted at assessment objective 2 (AO2). For example, you could be asked the question: 'Suggest why flooding may not happen despite the occurrence of very heavy rainfall in a drainage basin.' To score full marks, you must (i) apply geographical knowledge and understanding to this new context, and (ii) establish very clear connections between the question that is being asked and the stimulus material.

Your remaining short-answer questions will usually be purely knowledge-based, targeted at assessment objective 1 (AO1). They will be worth 4 or 5 marks and most likely use the command words 'explain', 'describe' or 'outline'. For example: 'Explain two ways in which vegetation can act as a water store.' High marks will be awarded to students who can write concise, detailed answers which incorporate and link together a range of geographical ideas, concepts or theories.

Eduqas
Short-answer questions 1 and 2 on your examination paper will be linked to figures (maps, charts, tables or photographs).

Part (a) of question 1 and part (a) of question 2 will always be targeted at AO3. This means that you must use geographical skills (AO3) to analyse or extract meaningful information or evidence from the figure. These questions will most likely use the command words 'describe', 'analyse' or 'compare'.

In part (b) of one of these questions, you may be asked to suggest possible reasons that could explain the information shown in the figure. This question will usually be worth 5 marks.

- This is called an applied knowledge task; it is targeted at AO2 and will most likely use the command word 'suggest'. An applied knowledge task will always include the instruction: 'Use the figure'.
- For example, two short questions could accompany a map showing soil infiltration rates recorded at different sites in a rural area with varying land uses, such as agriculture and forestry. The opening part (a) question could be: 'Analyse the pattern shown in the figure' (an AO3 task). The part (b) AO2 question which follows might ask: 'Suggest reasons why some sites have a lower infiltration rate than others shown in the figure.' To score full marks, you must (i) apply geographical knowledge and

understanding to this new context, and (ii) establish very clear connections between the question asked and the stimulus material.

The other part (b) question will usually be purely knowledge-based, targeted at AO1 and worth 5 marks. For example: 'Explain the differences between positive and negative feedback in systems.' High marks will be awarded to students who can write concise, detailed answers which incorporate and link together a range of geographical ideas, concepts or theories.

Evaluative essay writing

WJEC

You are given a choice of two 18-mark (10 marks AO1, 8 marks AO2) essays to write (*either* question 3 *or* question 4). These essays will most likely use the command words and phrases 'discuss', 'evaluate' or 'to what extent'. For instance:

- **Evaluate the relative importance of human activity as a cause of water cycle deficits.**
- **To what extent is human activity the most important cause of change in the size of global carbon stores?**

The box on page 217 provides guidance on how to answer these questions.

Eduqas

You are given a choice of two 20-mark (10 marks AO1, 10 marks AO2) essays to write (*either* question 3 *or* question 4). These essays will most likely use the command words and phrases 'discuss', 'evaluate' or 'to what extent'. These essays draw equally on *both* water cycle *and* carbon cycle knowledge. For instance:

- **'Precipitation plays an equally important role in both the water and carbon cycles.' Discuss.**
- **Discuss the severity of the impact of drought on different water and carbon stores.**

The box on page 217 provides guidance on how to answer these questions.

Synoptic geography

In addition to the three main AOs, some of your marks are awarded for 'synopticity'. The box on page 218 explains what this means.

The WJEC and Eduqas synoptic assessment

In the WJEC course, part of Unit 3 is devoted to synopticity while for Eduqas a similar assessment appears in Component 2. In both cases, synopticity is examined using an assessment called '21st Century Challenges'. This synoptic exercise consists of a linked series of four figures (maps, charts or photographs) with a choice of two accompanying essay questions. The WJEC question has a maximum mark of 26; for Eduqas it is 30. Possible questions include:

- **Discuss the view that physical processes (such as monsoon rainfall and overland flow) are mainly to blame for the accumulation of plastic in Earth's oceans.**
- **Physical processes can cause place identity to change rapidly whereas human activity always brings slower changes.**

As part of your answer, you will need to apply a range of knowledge from different topics, and also make good analytical use of the previously unseen resources in order to gain AO3 credit (the 'analysis and interpretation' features in this book have been carefully designed to help you in this respect). One or both of the questions may relate in part to the topic of Water and carbon cycles, as the example titles above demonstrate.

- Water cycle flows are an important reason why discarded plastic is washed in large volumes into the oceans each year, especially in monsoonal Asia. However, rising Asian affluence is a human factor that helps explain why far more plastic is being used (and then discarded). This essay allows you to make varied arguments using knowledge drawn from different parts of your geography specification.
- Note also how the second question allows you to explore water and carbon cycle flows (amongst other geography topics) operating on very different timescales. There is also an opportunity here to employ the Anthropocene concept which features in this book (pages 19–22 and 149–153).

Index

Acknowledgements

p. 54 Figure 2.21 River flow for the Coln (tributary of the Thames) at Bibury and the Thames at Kingston, London. Source: National River flow archive, Centre for Ecology, August 2017. Reprinted with permission; Figure 2.22 Average monthly discharge at selected stations in the Ganga-Brahmaputra-Meghna basin AND The causes of floods in the Ganges. Source: Hofer, T., and Messerli, B., 2006, Floods in Bangladesh, FAO and UN University Press. Reprinted with permission; **p. 74** Table 3.1 National selected water footprints; http://waterfootprint.org/en/water-footprint/national-water-footprint. Reprinted with permission of Water Footprint Network; Figure 3.10 Green-water, blue-water and grey-water footprints. Source: http://waterfootprint.org/media/downloads/Report50-NationalWaterFootprints-Vol1.pdf © 2011 M.M. Mekonnen and A.Y. Hoekstra. Reprinted with permission; **p. 90** Figure 3.20 The number of large dams worldwide, 1945 and 2005. Reprinted with permission of PennWell Corporation; **p. 109** Figure 4.7 Philosophical transactions. Series A, Mathematical, physical, and engineering sciences by Royal Society (Great Britain) Reproduced with permission of ROYAL SOCIETY in the format Republish in a book via Copyright Clearance Center; **p. 116** Figure 4.11, Figure 4.12 and Figure 4.13 Biogeosciences, (Thomas, M. V., Malhi, Y., Fenn, K. M., Fisher, J. B., Morecroft, M. D., Lloyd, C. R., Taylor, M. E., and McNeil, D. D.: Carbon dioxide fluxes over an ancient broadleaved deciduous woodland in southern England, Biogeosciences, 8, 1595-1613. This work is distributed under the Creative Commons Attribution 3.0 License; **p. 123** Figure 4.18 Olson, D., M., Terrestrial Ecoregions of the World: a new map of life on Earth, BioScience, Volume 51, Issue 11, 1 November 2001, Pages 933–938, by permission of Oxford University Press **p. 132** Table 5.2 Reprinted with permission of BP Statistical Review of World Energy 2017; Figure 5.5 Reprinted with permission of BP Statistical Review of World Energy 2016; **p. 133** Figure 5.6 Association for the Study of Peak Oil & Gas (ASPO); **p. 134** Table 5.3 Reprinted with permission of BP Statistical Review of World Energy 2017; **p. 135** Figure 5.7 Reprinted with permission of BP Statistical Review of World Energy 2017; **p. 136** Figure 5.8 and Figure 5.9 Reprinted with permission of BP Statistical Review of World Energy 2017; **p.150** Figure 5.19 Taylor, M. 2014 Global Warming and Climate Change. Chapter 2: Loading the dice: humans as planetary force. ANU Press. Article is covered under CC BY 4.0; **p.151** Figure 5.20 Philosophical transactions. Series A, Mathematical, physical, and engineering sciences by Royal Society (Great Britain) Reproduced with permission of ROYAL SOCIETY in the format Republish in a book via Copyright Clearance Center; **p.162** Table 6.1 Reprinted with permission of REN21, 2016, Capacity of renewable energy sources, 2004 and 2016, Renewables 2017 Global Status Report, Paris, REN21; Figure 6.2 Reprinted with permission of BP Statistical Review of World Energy 2016; **p. 166** Figure 6.6 Reprinted with permission of BP Statistical Review of World Energy 2016; **p. 172** Figure 6.10 Philosophical transactions. Series A, Mathematical, physical, and engineering sciences by Royal Society (Great Britain) Reproduced with permission of ROYAL SOCIETY in the format Republish in a book via Copyright Clearance Center; **p. 193** Figure 7.5 Nepstad, et al, 2008, p. 1738, THE ROYAL SOCIETY (U.K.) Article is covered under CC BY 4.0; **p. 194** Figure 7.6 Nepstad, et al, 2008, p. 1740, THE ROYAL SOCIETY (U.K.) Article is covered under CC BY 4.0; **p.195** Figure 7.7 Aragao et al., 2008, p.1781, THE ROYAL SOCIETY (U.K.) Article is covered under CC BY 4.0; **p. 199** Figure 7.10 Nepstad, et al, 2008, p. 1741, THE ROYAL SOCIETY (U.K.) Article is covered under CC BY 4.0; **p. 202** Figure 7.11 Oxford Research Encyclopedia of Environmental Science (OUP); **p. 203** Figure 7.12 World Wildlife Fund; **p. 208** Figure 7.16 Vegetation over Siberia (a) 2009 and (b) predicted for 2090. Source: Vygodskaya et al., 2007, Figure 4, p. 5 AND Figure 5, p.6 (http://iopscience.iop.org/article/10.1088/1748-9326/2/4/045033). Reprinted with permission.

We are grateful to Simon Oakes, editor of the *A-level Geography Topic Master* series, for inviting us to contribute to the series and for providing us with support and encouragement throughout. Our thanks to the team at Hodder for guiding this project through to completion, in particular Ruth Murphy and Margaret McGuire, who oversaw the editorial process for this book. Rob Bircher helped us to prepare and check the original manuscript, and Rhian McKay copyedited the text. Lydia Hardman was involved with copyediting and all proofreading stages, and arranged all the artworks, photos, permissions, and was instrumental in keeping the project on track. We are also indebted to Martin Evans, Professor of Geomorphology at the University of Manchester, who acted as academic reviewer for the book to ensure full accuracy of the contents: his detailed and insightful comments were invaluable and helped to materially improve the text.